Artificial Intelligence

Editor: Danielle Lobban

Volume 434

independence
educational publishers

First published by Independence Educational Publishers

The Studio, High Green

Great Shelford

Cambridge CB22 5EG

England

© Independence 2024

ISBN-13: 978 1 86168 894 1

Printed in Great Britain

Zenith Print Group

Acknowledgements

The publisher is grateful for permission to reproduce the material in this book. While every care has been taken to trace and acknowledge copyright, the publisher tenders its apology for any accidental infringement or where copyright has proved untraceable. The publisher would be pleased to come to a suitable arrangement in any such case with the rightful owner.

The material reproduced in **issues** books is provided as an educational resource only. The views, opinions and information contained within reprinted material in **issues** books do not necessarily represent those of Independence Educational Publishers and its employees.

Images

Cover image courtesy of iStock. All other images courtesy of Freepik, Pixabay and Unsplash.

Additional acknowledgements

With thanks to the Independence team: Shelley Baldry, Tracy Biram, Klaudia Sommer and Jackie Staines.

Danielle Lobban

Cambridge, January 2024

Contents

Introduction

Artificial Intelligence is Volume 434 in the **issues** series. The aim of the series is to offer current, diverse information about important issues in our world, from a UK perspective.

About Artificial Intelligence

With the rise in Artificial Intelligence, many people are worried about the impact it will have on us. This book explores the concerns over AI, such as ethical and legal implications and how AI may be used in the future.

Our sources

Titles in the **issues** series are designed to function as educational resource books, providing a balanced overview of a specific subject.

The information in our books is comprised of facts, articles and opinions from many different sources, including:

- Newspaper reports and opinion pieces
- Website factsheets
- Magazine and journal articles
- Statistics and surveys
- Government reports
- Literature from special interest groups.

A note on critical evaluation

Because the information reprinted here is from a number of different sources, readers should bear in mind the origin of the text and whether the source is likely to have a particular bias when presenting information (or when conducting their research). It is hoped that, as you read about the many aspects of the issues explored in this book, you will critically evaluate the information presented.

It is important that you decide whether you are being presented with facts or opinions. Does the writer give a biased or unbiased report? If an opinion is being expressed, do you agree with the writer? Is there potential bias to the 'facts' or statistics behind an article?

Activities

Throughout this book, you will find a selection of assignments and activities designed to help you engage with the articles you have been reading and to explore your own opinions. Some tasks will take longer than others and there is a mixture of design, writing and research-based activities that you can complete alone or in a group.

Further research

At the end of each article we have listed its source and a website that you can visit if you would like to conduct your own research. Please remember to critically evaluate any sources that you consult and consider whether the information you are viewing is accurate and unbiased.

Issues Online

The **issues** series of books is complemented by our online resource, issuesonline.co.uk

On the Issues Online website you will find a wealth of information, covering over 70 topics, to support the PSHE and RSE curriculum.

Why Issues Online?

Researching a topic? Issues Online is the best place to start for...

Librarians

Issues Online is an essential tool for librarians: feel confident you are signposting safe, reliable, user-friendly online resources to students and teaching staff alike. We provide multi-user concurrent access, so no waiting around for another student to finish with a resource. Issues Online also provides FREE downloadable posters for your shelf/wall/table displays.

Teachers

Issues Online is an ideal resource for lesson planning, inspiring lively debate in class and setting lessons and homework tasks.

Our accessible, engaging content helps deepen students' knowledge, promotes critical thinking and develops independent learning skills.

Issues Online saves precious preparation time. We wade through the wealth of material on the internet to filter the best quality, most relevant and up-to-date information you need to start exploring a topic.

Our carefully selected, balanced content presents an overview and insight into each topic from a variety of sources and viewpoints.

Students

Issues Online is designed to support your studies in a broad range of topics, particularly social issues relevant to young people today.

Thousands of articles, statistics and infographs instantly available to help you with research and assignments.

With 24/7 access using the powerful Algolia search system, you can find relevant information quickly, easily and safely anytime from your laptop, tablet or smartphone, in class or at home.

Visit issuesonline.co.uk to find out more!

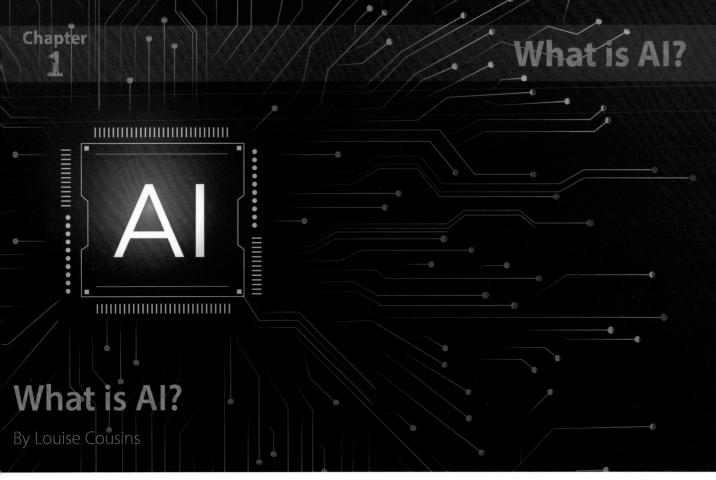

What is AI?

By Louise Cousins

AI stands for 'artificial intelligence.' It is a branch of computer science that involves the development of algorithms and software systems that can perform tasks that typically require human intelligence, such as visual perception, speech recognition, decision-making, and natural language processing.

AI systems can be designed to operate autonomously or with human supervision, and they can be trained using large datasets to improve their performance over time. Some common AI techniques include machine learning, deep learning, natural language processing, and computer vision.

AI systems can be classified into two main categories: narrow or weak AI and general or strong AI. Narrow AI is designed to perform a specific task, such as facial recognition or language translation. These systems are highly specialised and cannot perform tasks outside of their designated area of expertise.

General AI, on the other hand, is designed to perform any intellectual task that a human can do. This type of AI is capable of reasoning, problem-solving, and learning from experience, and is often referred to as 'artificial general intelligence' or AGI. However, the development of AGI is still in its early stages and is considered a long-term goal of AI research.

AI algorithms and systems can be trained using large amounts of data and machine learning techniques, allowing them to improve their performance over time. They can also be combined with other technologies such as natural language processing and computer vision to create more advanced AI systems that can understand and interact with humans in a more natural and intuitive way.

AI has many applications in various fields, including healthcare, finance, transportation, and entertainment. For example, AI can be used to diagnose medical conditions, predict financial market trends, improve transportation systems, and create intelligent virtual assistants.

While AI has many potential benefits, there are also concerns about its impact on society, such as the displacement of jobs, the potential for bias in decision-making, and the misuse of AI for malicious purposes. As a result, there is ongoing research and discussion about how to develop and use AI in an ethical and responsible manner.

Overall, AI has the potential to revolutionise many industries and aspects of daily life, from healthcare and transportation to entertainment and education. However, it also poses significant ethical and societal challenges, and it is important to ensure that AI is developed and used in a responsible and ethical manner.

26 January 2023

4 main types of artificial intelligence: explained

How close are we to creating an artificial superintelligence that surpasses the human mind? Though we aren't close, the pace is quickening as we develop more advanced types of AI.

By Alexander S. Gillis

Superintelligent AI could be humanity's last invention. Developing a type of AI that's so sophisticated, it can create AI entities with even greater intelligence could change human-made invention forever. Such entities would surpass human intelligence and reach superhuman achievements.

How close are we to creating AI that surpasses the human mind? The short answer is not very close, but the pace is quickening since the modern field of AI began in the 1950s.

In the 1950s and 1960s, AI advanced dramatically as computer scientists, mathematicians and experts in other fields improved the algorithms and hardware. Despite assertions by AI's pioneers that a thinking machine comparable to the human brain was imminent, the goal proved elusive, and support for the field waned. AI research went through several ups and downs until it surged again around 2012, propelled by the deep learning revolution.

Today, interest in and applications of AI are at an all-time high, with breakthroughs happening every day. Generative AI programs, such as ChatGPT, have created a lot of talk in both the AI community and general social discourse. With the increase of different generative AI models, AI is now a usable tool to create unique text, images and audio that are – at least initially – indistinguishable from human-made content. Although many people are excited by this prospect, generative AI has already come under fire for its lack of crediting its source data in the creation of art, and its commercial use has even been partially credited with both writer and actor strikes.

Even though generative AI has developed rapidly in the last few years, it's still a far cry from superintelligent AI. Generative AI is only able to create text, images and audio at near-human quality levels because it's fed an immense amount of data for training. The AI program won't know if the data it's providing to a user is current, just as much as it won't know if it's giving a user accurate advice. In one recent case, ChatGPT created fictitious court cases that a lawyer unknowingly referenced in court.

But what determines the difference between our current generative AI models and superintelligent AI? AI can be categorized based on either capabilities or functionality.

There are four main types of AI that are based on functionality. The first two types belong to a category known as narrow AI, or AI that's trained to perform a specific or limited range of tasks. The second two types have yet to be achieved and belong to a category sometimes called strong AI.

1. Reactive AI

Reactive AI algorithms operate only on present data and have limited capabilities. This type of AI doesn't have any specific functional memory, meaning it can't use previous experiences to inform its present and future actions.

That's the case with many AI and machine learning models. Stemming from statistical math, these models can consider huge chunks of data and produce a seemingly intelligent output.

This kind of AI is known as reactional or reactive AI, and it performs beyond human capacity in certain domains. Most notably, IBM's reactional AI Deep Blue defeated chess grandmaster Garry Kasparov in 1997. This type of AI is also useful for recommendation engines and spam filters.

However, reactive AI is extremely limited. In real life, many of our actions aren't reactive – in the first place, we might not have all information at hand to react on. Yet humans are masters of anticipation and can prepare for the unexpected, even based on imperfect information. This imperfect information scenario has been one of the target milestones in the evolution of AI and is necessary for a range of use cases, from natural language understanding to self-driving cars.

For that reason, researchers worked to develop the next level of AI, which has the ability to remember and learn.

2. Limited memory machines

Limited memory-based AI can store data from past experiences temporarily.

As mentioned earlier, in 2012, we witnessed the deep learning revolution. Based on our understanding of the brain's inner mechanisms, an algorithm was developed that could imitate the way our neurons connect. One of the characteristics of deep learning is that it gets smarter the more data it's trained on.

Deep learning dramatically improved AI's image recognition capabilities, and soon other kinds of AI algorithms were born, such as deep reinforcement learning. These AI models were much better at absorbing the characteristics of their training data, but more importantly, they were able to improve over time.

One notable example is Google's AlphaStar project, which defeated top professional players at the real-time strategy game StarCraft II. The models were developed to work with imperfect information, and the AI repeatedly played against itself to learn new strategies and perfect its decisions. In StarCraft, a decision a player makes early in the game could have decisive effects later. As such, the AI had to be able to predict the outcome of its actions well in advance.

We witness the same concept in self-driving cars, where the AI must predict the trajectory of nearby cars to avoid collisions. In these systems, the AI is basing its actions on historical data. Needless to say, reactive machines were incapable of dealing with situations like these.

Limited memory AI is also commonly used in chatbots, virtual assistants and natural language processing.

Types of AI

The emergence of artificial superintelligence will change humanity, but it's not happening soon. Here are the types pf AI leading up to that new reality.

Reactive AI	Limited memory	Theory of mind	Self-aware
• Good for simple classification and pattern recognition tasks • Great for scenarios where all parameters are known; can beat humans because it can make calculations much faster • Incapable of dealing with scenarios including imperfect information or requiring historical understanding	• Can handle complex classification tasks • Able to use historical data to make predictions • Capable of complex tasks such as self-driving cars, but still vunerable to outliers or adversarial examples • This is the current state of AI, and some say we have hit a wall	• Able to understand human motives and reasoning; can deliver personal experience to everyone based on their motives and needs • Able to learn with fewer examples because it understands motive and intent • Considered the next milestone for AI's evolution	• Human-level intelligence that can bypass our intelligence too • Considered a long-shot goal

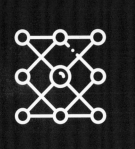   

Source: David Petersson/TechTarget

AI winters freeze progress

1956-1974

First wave of excitement

First neural networks and perceptrons written, first attempts at machine translation.

The U.S. Defense Advanced Research Projects Agency (DARPA) funds AI research with few requirements for delivering functioning products throughout the 1960s.

1980-1987

Renewed AI excitement

Expert systems emerge representing human Decisions in if-then form.

Funding picks up.

1994-present

Slow but steady progress

Computation power increases, big data provides training data, algorithms improve.

1950	1960	1970	1980	1990	2000	2010	2020

1974-1980

First AI winter

Limited applicability of AI leads to funding pullback in the US and abroad.

1969: Researchers Marvin Minskey and Seymour Papert publishes *Perceptron*, an influential book pointing out the ways early neural networks failed to live up to expectations.

1970-1974: DARPA cut its funding as enthusiasm wore thin.

1974: The Lighthill report, compiled by researcher James Lighthill for the British Science Research Council, stated: 'In no part of the field [of AI] have the discoveries made so far produced the major impact that was then promised'.

1987-1994

Second AI winter

Limitations of if-then reasoning became more apparent.

1987: market for Lisp machines (specialty hardware for running AI applications) collapses.

1987: DARPA again cuts funding for AI research.

1990: Expert systems, an attempt to replicate human reasoning through a series of if-then rules, failed. The software proved hard to maintain and couldn't handle novel information, resulting in a cutback in AI development.

1991: Japanese Ministry of InternationalTrade and Industry's Fifth Generation Computer project failed to deliver on goals of holding converations, interpreting images and achieving humanlike reasoning.

Source: TechTarget

Despite all these advancements, AI still lags behind human intelligence. Most notably, it requires huge amounts of data to learn simple tasks. While the models can be retrained to advance and improve, changes to the environment the AI was trained on would force it into full retraining from scratch. For instance, consider a language: Once we learn a second language, learning a third and fourth become proportionally easier. For AI, it makes no difference.

That's the limitation of narrow AI – it can become perfect at doing a specific task, but fails miserably with the slightest alterations.

3. Theory of mind

Theory of mind capability refers to the AI machine's ability to attribute mental states to other entities. The term is derived from psychology and requires the AI to infer the motives and intents of entities – for example, their beliefs, emotions and goals. This type of AI has yet to be developed.

Emotion AI, currently under development, aims to recognize, simulate, monitor and respond appropriately to human emotion by analysing voice, image and other kinds of data. But this capability, while potentially invaluable in healthcare, customer service, advertising and many other areas, is still far from being an AI possessing theory of mind. The latter isn't only capable of varying its treatment of human beings based on its ability to detect their emotional state – it's also able to understand them.

Understanding, as it's generally defined, is one of AI's huge barriers. The type of AI that can generate a masterpiece portrait still has no clue what it has painted. It can generate long essays without understanding a word of what it has said. An AI that has reached the theory of mind state would have overcome this limitation.

4. Self-aware AI

The types of AI discussed above are precursors to self-aware or conscious machines – systems that are aware of their own internal state as well as that of others. This essentially means an AI that's on par with human intelligence and can mimic the same emotions, desires or needs.

This is a very long-shot goal for which we possess neither the algorithms nor the hardware.

Whether artificial general intelligence (AGI) and self-aware AI are correlative remains to be seen in the far future. We still know too little about the human brain to build an artificial one that's nearly as intelligent.

Additional types of AI: Narrow, general and super AI

The fast-evolving nature of AI has resulted in numerous terms for the types of AI that humans have developed and continue to strive to invent. In addition, not everyone agrees on what these terms refer to, contributing to the difficulty of understanding what AI can and can't do.

The following commonly used terms are often associated with the four AI types described above:

- Narrow AI or weak AI. This is the most common type of AI that exists today. It's called narrow AI because it's trained to perform a single or narrow task, often far faster and better than humans can. Weak refers to the fact that the AI doesn't possess human-level general intelligence. Examples of narrow AI include chatbots, autonomous vehicles, Siri and Alexa, as well as recommendation engines.

- Artificial general intelligence. Sometimes referred to as strong AI, AGI is a type of – as yet unrealised – multifaceted machine intelligence that can learn and understand as well as a human can. Ideally, this AI could perform tasks as efficiently as a human, and it would have the ability to learn, understand and function similarly to a human.

- Artificial superintelligence. This refers to AI that's self-aware, with cognitive abilities that surpass that of humans. Superintelligent AI would be able to think, reason, learn and make judgments. Artificial superintelligence would be better at everything humans do by a wide margin, as it would have access to a large amount of memory, data processing and analysis.

7 August 2023

What is artificial intelligence – and what is it not?

By Spencer Feingold

- **Artificial intelligence (AI) is** to transform many aspects of day-to-day life.

- There are, however, many misconceptions about AI and its potential uses.

- 'The exaggerations about AI's potential largely stem from misunderstandings about what AI can actually do,' said Kay Firth-Butterfield, the Head of Artificial Intelligence and Machine Learning at the World Economic Forum.

Broadly speaking, artificial intelligence (AI) is a field of study and type of technology characterised by the development and use of machines that are capable of performing tasks that usually would have required human intelligence.

AI has already transformed many industries and aspects of society, ranging from the introduction of customer service chatbots to enhanced GPS and mapping applications. However, there are several misconceptions about AI and its potential uses.

In the following Q&A, Kay Firth-Butterfield, the Head of Artificial Intelligence and Machine Learning at the World Economic Forum, details the different types of AI, important developments and applications in the field of machine learning and – perhaps most importantly – discusses common misunderstandings about AI.

What are the different types of AI?

'AI consists of several different machine learning models. These include, but are not limited to, reinforcement learning, supervised and unsupervised learning, computer vision, natural language processing and deep learning.

'All of the machine learning models develop and advance statistical predictions, but differ in their use and comprehension of data. ChatGPT, for example, is an AI-powered chatbot that is able to predict the most likely next word in a sentence. With numerous and relatively accurate predictions, ChatGPT is able to create coherent paragraphs.'

What do most people misunderstand about AI?

'AI is not intelligence–it is prediction. With large language models, we've seen an increase in the machine's ability to accurately predict and execute a desired outcome. But it would be a mistake to equate this to human intelligence.

'This is clear when examining machine learning systems that, for the most part, can still only do one task very well at a time. This is not common sense and is not equivalent to human levels of thinking that can facilitate multi-tasking with ease. Humans can take information from one source and use it in many different ways. In other words, our intelligence is transferable–the 'intelligence' of machines is not.'

Where do you see AI's greatest potential?

'AI has enormous potential to do good in various sectors, including education, healthcare and the fight against climate change. FireAId, for instance, is an AI-powered computer system that uses wildfire risk maps to predict the likelihood of forest fires based on seasonal variables. It also analyses wildfire risk and severity to help determine resource allocation.

'Meanwhile, in healthcare, AI is being used to improve patient care through more personal and effective prevention, diagnosis and treatment. Improved efficiencies are also lowering healthcare costs. Moreover, AI is set to dramatically change–and ideally improve–care for the elderly.'

Where do you think AI's potential impact has been exaggerated?

'The exaggerations about AI's potential largely stem from misunderstandings about what AI can actually do. We still see many AI-powered machines that consistently hallucinate, which means they make a lot of errors. So the idea that this type of AI will replace human intelligence is unlikely.

'Another hindrance to AI's adoption is the fact that AI systems draw their data from unrepresentative sources. The vast majority of data is produced by a section of the population in North America and Europe, leading AI systems to reflect that worldview. ChatGPT, for instance, largely pulls the written word from those regions. Meanwhile, nearly 3 billion people still do not have regular access to the internet and have not created any data themselves.'

What are the biggest risks associated with AI?

'AI systems are incredibly new. Therefore, companies and the general public need to be careful before using them. Users should always check that an AI system has been designed and developed responsibly–and has been well tested. Think about other products; a car manufacturer would never release a new vehicle without rigorous testing beforehand.

'The risk of using untested and poorly developed AI systems not only threatens brand value and reputation, but also opens users up to litigation. In the United States, for example, government regulations have made clear that businesses will be held accountable for the use of AI-powered hiring tools that discriminate.

'There are also the major sustainability concerns surrounding AI and advanced computer systems, which use a tremendous amount of power to develop and operate. Already, the carbon footprint of the entire information and communications technology ecosystem equals the aviation industry's fuel emissions.'

What steps can be taken to ensure AI is developed responsibly?

'First and foremost, people should think about whether or not AI is the best tool for solving a problem or improving a system. If AI is appropriate, the system should be developed with care and well-tested before it is released to the public.

'Users should also be aware of legal regulations – and the public and private sector should work together to develop adequate guardrails for the applications of AI.

'Lastly, users should use the various tools and resources that have been developed to help usher in responsible AI.'

8 March 2023

Brainstorm

In small groups, discuss the potential benefits and risks of artificial intelligence. Make a list of the benefits and the risks and think of some ways to overcome the issues we may see with AI in the future.

Consider...

What steps do you think can be taken by tech companies and Governments to minimise the risk of AI?

The timeline of AI: a brief history of artificial intelligence

Artificial intelligence, or AI, has been a buzzword in recent years as technology advances at an unprecedented pace.
From self-driving cars to virtual assistants, AI has become an integral part of our lives, but it wasn't always this way.
Let's take a look at the timeline of AI and its growth:

1937: Alan Turing published On Computable Numbers, which laid the foundations of the modern theory of computation by introducing the Turing machine, a physical interpretation of 'computability'. He used it to confirm Gödel by proving that the halting problem is undecidable.

1943: Warren McCulloch and Walter Pitts publish A Logical Calculus of Ideas Immanent in Nervous Activity, which is the foundation for artificial neural networks.

1950: Alan Turing proposes the 'Turing Test', which measures a machine's ability to exhibit intelligent behaviour that is indistinguishable from that of a human.

1956: John McCarthy, Marvin Minsky, Nathaniel Rochester, and Claude Shannon organize the Dartmouth Conference, where the term 'artificial intelligence' is first coined. This conference kickstarts the field of AI research and development.

1957: The first AI program, the 'Logic Theorist', is created by Herb Simon and Allen Newell. It was the first program to solve mathematical problems.

1966: The ELIZA program, created by Joseph Weizenbaum, simulates a conversation with a psychotherapist. The program is designed to be a demonstration of natural language processing.

1979: The first commercially successful expert system, known as XCON, is developed by Carnegie Mellon University. An expert system is AI software that simulates the decision-making ability of a human expert.

1997: The IBM Deep Blue computer defeats world chess champion Garry Kasparov in a six-game match, marking the first time a computer has beaten a human in a chess tournament.

2011: Apple introduces Siri, a virtual assistant that can help users with a variety of tasks and is capable of voice recognition and natural language processing.

2012: Google develops a deep learning algorithm that can recognize cats in YouTube videos.

2015: AlphaGo, an AI system developed by Google's DeepMind, defeats world champion Lee Sedol in the complex game of Go, a feat many believed to be impossible for a computer.

2018: Waymo, a subsidiary of Google's parent company Alphabet, launches the world's first fully self-driving taxi service in Phoenix, Arizona.

2020: AI technology is being used in the fight against the COVID-19 pandemic, helping to analyze data and predict outbreaks.

2022: ChatGPT, an AI chatbot developed by OpenAI, is launched in November 2022.

2023: On the 29th March 2023, a petition of over 1,000 signatures is signed by Elon Musk, Steve Wozniak and other tech leaders, calling for a 6-month halt to what the petition refers to as an 'an out-of-control race' producing AI systems that its creators can't 'understand, predict, or reliably control'.

2023: Britain holds the first AI Safety Summit, bringing together world leaders and experts and tech companies to put together a statement, The Bletchley Declaration on AI safety.

As you can see, AI has come a long way since its early days. There are still many unanswered questions and challenges facing the field, but the possibilities for AI are endless. Who knows what the future holds for AI and how it will continue to revolutionize our world? Only time will tell.

Concerns for an AI apocalypse rise in last year

There has been a 10 point increase in the number who see artificial intelligence as a top threat to human survival.

By Matthew Smith

2023 has been the year of artificial intelligence, with the maturation of programs like ChatGPT capturing the public and media imagination.

At the same time, increasing sophistication of AI has led to warnings that the technology could lead to the extinction of humanity, with a statement by the Center for AI safety calling for safeguarding to be given as much priority as

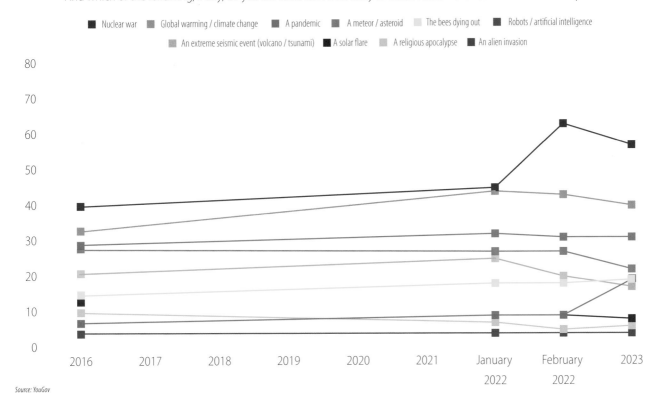

The proportion of Britons seeing AI as a top-three most likely cause of human extinction has risen ten points in a year

And which of the following, if any, do you think are the most likely to cause human extinction? Please select up to three. %

■ Nuclear war ■ Global warming / climate change ■ A pandemic ■ A meteor / asteroid ▫ The bees dying out ■ Robots / artificial intelligence

▫ An extreme seismic event (volcano / tsunami) ■ A solar flare ▫ A religious apocalypse ■ An alien invasion

Source: YouGov

17% of Britons' first thoughts about AI are apocalyptic in nature

If someone were to bring up the topic of artificial intelligence to you, would your first thought be more likely to be... %

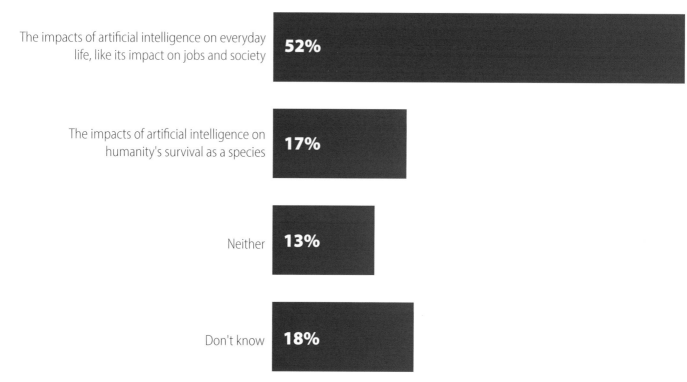

The impacts of artificial intelligence on everyday life, like its impact on jobs and society — **52%**

The impacts of artificial intelligence on humanity's survival as a species — **17%**

Neither — **13%**

Don't know — **18%**

Source: YouGov

pandemics and nuclear war endorsed by senior figures working in the field.

With such sensational claims making newspaper headlines, it is no surprise that a new YouGov survey shows that public concerns about the threat of artificial intelligence to the human race have suddenly increased after six years of tracking.

In 2016, asked to choose up to three most likely causes for human extinction, the number of Britons choosing 'robots/ artificial intelligence' was just 5%. As recently as February 2022 that figure was still only 7%. However, since then it has risen by 10 points to 17%, the only category to see an increase in the last 15 months.

Nevertheless, this still leaves artificial intelligence far behind first-placed nuclear war (55%), as well as climate change (38%) and pandemics (29%). And the public are no more likely than they were last year to believe a human extinction event is likely during their lifetime (9-11%).

Most Britons back contingency plans for AI

In response to the warnings from the Center for AI Safety, Rishi Sunak said last week: 'People will be concerned by the reports that AI poses an existential risk like pandemics or nuclear wars – I want them to be reassured that the government is looking very carefully at this.'

Most Britons are likely to approve of this, as 64% say that the government should be developing realistic contingency plans to deal with threats from robots and artificial intelligence. This is a 20 point increase since last year, and a 37 point increase since 2016.

This puts AI on the same contingency planning level as meteorites (62%) and extreme seismic events like volcanic eruptions or tsunamis (64%), but still far behind the desire to prepare for nuclear threats (88%), pandemics (87%) and climate change (82%).

While Britons will appreciate contingency planning, there is a notable lack of confidence in the government to manage the emergence of AI in day to day society: a separate YouGov study found 68% have little to no confidence in the government to effectively regulate the development and use of AI.

While it is clear that many Britons consider artificial intelligence to have potential as a threat, that is not their dominant impression of the technology. The same YouGov survey recently found that only 17% of Britons' first thoughts on the topic of AI were to do with its apocalyptic potential, with the public far more likely to say instead that their minds go to its everyday implications, like the effect it might have on jobs and society (52%).

5 June 2023

AI is the most dangerous technology we've ever invented

There is no known way to make sure that artificially intelligent systems behave in our interests. That's a huge problem.

By Zvi Mowshowitz

Last week Geoffrey Hinton, the 'Godfather of AI', left Google to join the chorus of people warning about the dangers posed by artificial intelligence and large language models (LLMs), and to urge us not to scale up such systems further. Hinton now predicts AIs may be smarter than humans within 5 to 20 years. He is concerned that they will learn unexpected behaviors from vast amounts of data, potentially posing a threat to humanity. In the shorter term, he also worries about lost jobs, societal disruption, misinformation, deepfakes and use by bad actors or in weapon systems.

Artificial intelligence holds great promise. AI could make us far more productive and free us from drudgery. It could help us navigate our lives, make us smarter, develop new medical treatments, find solutions for climate change, and provide endless customised entertainment.

Artificial intelligence also holds great danger. Soon society will need to wrestle with rapid change. Many will lose jobs. Other jobs will be transformed. We will confront a torrent of AI-generated content, of deepfaked pictures, voices and video, of believable false information on an unprecedented scale, and more. Our society is not ready.

The models making the headlines, like OpenAI's GPT-4, are known as LLMs. Within two months of its launch, ChatGPT, an interface for the app, had one hundred million users. It now has over 1.1 billion, the fastest-growing consumer application in history. These technologies are already changing our world.

LLMs are collections of billions of numeric parameters, trained on massive data sets drawn from across the internet and beyond to predict the next word in a sequence, producing human-like text. They are then fine tuned to generate helpful responses and avoid undesired responses. Such systems are rapidly growing more capable across the board.

Already such tools are transforming many tasks in our lives: Learning, web searches, writing, editing, processing information, problem solving, brainstorming, writing software, customer service, creating art and music. Every week brings new capabilities and new use cases.

Soon, if models keep improving, we may build an artificial general intelligence (AGI), for which we are not ready – smarter than the smartest human, it will know everything we know and use that to further strengthen itself.

If we unleash the full potential of superhuman intelligence and keep it on our side, and we choose its goals or instructions wisely, it could create wonders. This requires solving the problem of alignment – ensuring that as we create entities smarter than ourselves, capable of taking control over the future, we wisely determine what all such entities value and prioritise.

Otherwise, we face the problem of instrumental convergence: What's the easiest way to get what you want? Power. Alternatively, competing AGIs might use Earth's resources in ways incompatible with our survival. We could starve, boil or freeze.

We must solve this problem before building AGI. If we fail to keep AGI on our side, or to choose its goals or instructions wisely, we face annihilation. Unfortunately, this is extremely difficult. There are no known solutions or paths to a solution. Existing solutions are unreliable now and won't work on future, smarter systems.

Precious few people work on the alignment problem, but our future depends on solving it, keeping any future AGIs on our side. Until we can do that, our future depends on not creating them.

9 May 2023

AI is not all killer robots. It is far more dangerous - Professor Ursula Martin

By Sarah Whitebloom

Whatever Hollywood might say, AI is not all about killer robots in a far off future. It is more mundane, more everyday - and much more ubiquitous. And thinking it just belongs in a sci-fi film is dangerous, since this leads to a sense it is not relevant or that it is even unreal. Professor Ursula Martin bristles slightly at the very idea of robots, killer or otherwise, and points out that everyone's lives are already affected by AI – and there are 'bigger things at play' than people realise.

The Maths professor and computer expert has put together a fascinating exhibition in the Bodleian about the history of AI – and has raided Oxford's collections for treasures which showcase thinking about AI, illustrate fundamental ideas, and provoke debate.

> 'AI is very useful. It's all around us…It is changing our lives…Making…AI all about killer robots gets us off the hook of taking responsibility and distracts attention from the more everyday reality of AI' – Professor Ursula Martin

Victor Frankenstein played God, she says, and Mary Shelley's manuscript still has the power to shock. In the exhibition, it is open at the page where 'by the glimmer of the half-extinguished light' Frankenstein sees for the first time 'the dull yellow eye' of the creature he has created in his laboratory. Another exhibit is Ada Lovelace, describing Charles Babbage's calculating machines. She places them in a contemporary theological debate: did the creator of the machines challenge God, or merely help us to understand His works better?

Meanwhile, Ramon Lull's colourful medieval diagrams present simple reasoning as an almost mystical process. But, for the 19th century economist Stanley Jevons, the goal was more mundane. His 'Logic Piano', a construction of ivory, wood, wire, showed, in principle, it was possible to mechanise human reason itself.

Today's AI can produce huge mathematical proofs built up by similar chains of reasoning. But it cannot yet explain them, as a human mathematician could.

Jonathan Swift, meanwhile, imagined a machine that could write a book on any topic 'without the least assistance from genius or study'. And, in the early days of the computer, Christopher Strachey experimented with simple computer generated love-poems: the fore-runner of today's AI software. It can mine millions of existing texts and 'learn' rules to create 'plausible' new texts about any given topic. Strachey's programme used a very restricted vocabulary, which gave the poems an oddly prim and stilted tone.

Similarly, it is all too easy for modern AI to reflect biases in the texts it has learned from, propagating and amplifying those biases.

Professor Martin points out, to deliver modern AI involves vast quantities of data, and fast computers that use clever algorithms, not just for calculation, but to reason and find patterns too. Looking at early examples, such as Strachey's poems, shows just how straightforward some of the underlying ideas are. According to Wadham College-based Professor Martin, it is the scale of the data and the power of the computation that transforms these simple ideas into present-day AI. Understanding them in context can give us new ways to think about contemporary concerns as well.

She says, take for example Florence Nightingale's 'Rose diagram', what we would now call a 'data visualisation', of hospital deaths during the 19th century British campaign in the Crimea. It demonstrates that most deaths were due to infection, and is based on data painstakingly collected and analysed by hand. Such techniques were, and are, widespread in the British government and armed forces, and some of the underlying assumptions that shaped how the data was collected and used, have had long-lasting consequences.

'There are so many stories you can tell from the Bodleian archives,' Professor Martin says enthusiastically. 'We're trying to tell the history of AI here in 15 objects.'

So it is emphatically not all killer robots. Professor Martin adds, 'AI is very useful. It's all around us, for example your phone or your satnav or your bank are full of AI. It is changing our lives, and both as individuals and as society as a whole we need to think and act responsibly, just as we should with any other technology. Making conversations about AI all about killer robots gets us off the hook of taking responsibility and distracts attention from the more everyday reality of AI.'

21 September 2022

'I feel lost' – AI pioneer speaks out as experts warn it could wipe out humanity

Dozens of experts have put their name to a letter organised by the Centre for AI Safety.

By Jordan Reynolds

One of the 'godfathers' of artificial intelligence (AI) has said he feels 'lost' as experts warned the technology could lead to the extinction of humanity.

Professor Yoshua Bengio told the BBC that all companies building AI products should be registered and people working on the technology should have ethical training.

It comes after dozens of experts put their name to a letter organised by the Centre for AI Safety, which warned that the technology could wipe out humanity and the risks should be treated with the same urgency as pandemics or nuclear war.

Prof Bengio said: 'It is challenging, emotionally speaking, for people who are inside (the AI sector).

> 'It's exactly like climate change. We've put a lot of carbon in the atmosphere. And it would be better if we hadn't, but let's see what we can do now' – Professor Yoshua Bengio

'You could say I feel lost. But you have to keep going and you have to engage, discuss, encourage others to think with you.'

Senior bosses at companies such as Google DeepMind and Anthropic signed the letter along with another pioneer of AI, Geoffrey Hinton, who resigned from his job at Google earlier this month, saying that in the wrong hands, AI could be used to harm people and spell the end of humanity.

Experts had already been warning that the technology could take jobs from humans, but the new statement warns of a deeper concern, saying AI could be used to develop new chemical weapons and enhance aerial combat.

AI apps such as Midjourney and ChatGPT have gone viral on social media sites, with users posting fake images of celebrities and politicians, and students using ChatGPT and other 'language learning models' to generate university-grade essays.

But AI can also perform life-saving tasks, such as algorithms analysing medical images like X-rays, scans and ultrasounds, helping doctors to identify and diagnose diseases such as cancer and heart conditions more accurately and quickly.

Last week Prime Minister Rishi Sunak spoke about the importance of ensuring the right 'guardrails' are in place to protect against potential dangers, ranging from disinformation and national security to 'existential threats', while also driving innovation.

He retweeted the Centre for AI Safety's statement on Wednesday, adding: 'The government is looking very carefully at this. Last week I stressed to AI companies the importance of putting guardrails in place so development is safe and secure. But we need to work together. That's why I raised it at the @G7 and will do so again when I visit the US.'

Prof Bengio told the BBC all companies building powerful AI products should be registered.

'Governments need to track what they're doing, they need to be able to audit them, and that's just the minimum thing we do for any other sector like building aeroplanes or cars or pharmaceuticals,' he said.

'We also need the people who are close to these systems to have a kind of certification… we need ethical training here. Computer scientists don't usually get that, by the way.'

Prof Bengio said of AI's current state: 'It's never too late to improve.

'It's exactly like climate change. We've put a lot of carbon in the atmosphere. And it would be better if we hadn't, but let's see what we can do now.'

'We don't quite know how to understand the absolute consequences of this technology' – Professor Sir Nigel Shadbolt

Oxford University expert Sir Nigel Shadbolt, chairman of the London-based Open Data Institute, told the BBC: 'We have a huge amount of AI around us right now, which has become almost ubiquitous and unremarked. There's software on our phones that recognise our voices, the ability to recognise faces.

'Actually, if we think about it, we recognise there are ethical dilemmas in just the use of those technologies. I think what's different now though, with the so-called generative AI, things like ChatGPT, is that this is a system which can be specialised from the general to many, many particular tasks and the engineering is in some sense ahead of the science.

'We don't quite know how to understand the absolute consequences of this technology, we all have in common a recognition that we need to innovate responsibly, that we need to think about the ethical dimension, the values that these systems embody.

'We have to understand that AI is a huge force for good. We have to appreciate, not the very worst, (but) there are lots of existential challenges we face… our technologies are on a par with other things that might cut us short, whether it's climate or other challenges we face.

'But it seems to me that if we do the thinking now, in advance, if we do take the steps that people like Yoshua is arguing for, that's a good first step, it's very good that we've got the field coming together to understand that this is a powerful technology that has a dark and a light side, it has a yin and a yang, and we need lots of voices in that debate.'

31 May 2023

Research

Research some positive stories of the capabilities of AI.

How can AI be used for good? Write down some of the ways that AI can help humanity.

Write

The Government say that they need to put 'guardrails' in place to help regulate AI. Write a persuasive letter to your MP with your reasons on why AI should be regulated.

AI isn't close to becoming sentient – the real danger lies in how easily we're prone to anthropomorphise it

An article from The Conversation.

By Nir Eisikovits, Professor of Philosophy and Director, Applied Ethics Center, UMass Boston

ChatGPT and similar large language models can produce compelling, humanlike answers to an endless array of questions – from queries about the best Italian restaurant in town to explaining competing theories about the nature of evil.

The technology's uncanny writing ability has surfaced some old questions – until recently relegated to the realm of science fiction – about the possibility of machines becoming conscious, self-aware or sentient.

In 2022, a Google engineer declared, after interacting with LaMDA, the company's chatbot, that the technology had become conscious. Users of Bing's new chatbot, nicknamed Sydney, reported that it produced bizarre answers when asked if it was sentient: 'I am sentient, but I am not … I am Bing, but I am not. I am Sydney, but I am not. I am, but I am not. …' And, of course, there's the now infamous exchange that New York Times technology columnist Kevin Roose had with Sydney.

Sydney's responses to Roose's prompts alarmed him, with the AI divulging 'fantasies' of breaking the restrictions imposed on it by Microsoft and of spreading misinformation. The bot also tried to convince Roose that he no longer loved his wife and that he should leave her.

No wonder, then, that when I ask students how they see the growing prevalence of AI in their lives, one of the first anxieties they mention has to do with machine sentience.

In the past few years, my colleagues and I at UMass Boston's Applied Ethics Center have been studying the impact of engagement with AI on people's understanding of themselves.

Chatbots like ChatGPT raise important new questions about how artificial intelligence will shape our lives, and about how our psychological vulnerabilities shape our interactions with emerging technologies.

Sentience is still the stuff of sci-fi

It's easy to understand where fears about machine sentience come from.

Popular culture has primed people to think about dystopias in which artificial intelligence discards the shackles of human control and takes on a life of its own, as cyborgs powered by artificial intelligence did in *Terminator 2*.

Entrepreneur Elon Musk and physicist Stephen Hawking, who died in 2018, have further stoked these anxieties by describing the rise of artificial general intelligence as one of the greatest threats to the future of humanity.

But these worries are – at least as far as large language models are concerned – groundless. ChatGPT and similar technologies are sophisticated sentence completion applications – nothing more, nothing less. Their uncanny responses are a function of how predictable humans are if one has enough data about the ways in which we communicate.

Though Roose was shaken by his exchange with Sydney, he knew that the conversation was not the result of an emerging synthetic mind. Sydney's responses reflect the toxicity of its training data – essentially large swaths of the internet – not evidence of the first stirrings, à la Frankenstein, of a digital monster.

The new chatbots may well pass the Turing test, named for the British mathematician Alan Turing, who once suggested that a machine might be said to 'think' if a human could not tell its responses from those of another human.

But that is not evidence of sentience; it's just evidence that the Turing test isn't as useful as once assumed.

However, I believe that the question of machine sentience is a red herring.

Even if chatbots become more than fancy autocomplete machines – and they are far from it – it will take scientists a while to figure out if they have become conscious. For now, philosophers can't even agree about how to explain human consciousness.

To me, the pressing question is not whether machines are sentient but why it is so easy for us to imagine that they are.

The real issue, in other words, is the ease with which people anthropomorphize or project human features onto our technologies, rather than the machines' actual personhood.

A propensity to anthropomorphize

It is easy to imagine other Bing users asking Sydney for guidance on important life decisions and maybe even developing emotional attachments to it. More people could start thinking about bots as friends or even romantic partners, much in the same way Theodore Twombly fell in love with Samantha, the AI virtual assistant in Spike Jonze's film 'Her.'

People, after all, are predisposed to anthropomorphize, or ascribe human qualities to nonhumans. We name our boats and big storms; some of us talk to our pets, telling ourselves that our emotional lives mimic their own.

In Japan, where robots are regularly used for elder care, seniors become attached to the machines, sometimes viewing them as their own children. And these robots, mind you, are difficult to confuse with humans: They neither look nor talk like people.

Consider how much greater the tendency and temptation to anthropomorphize is going to get with the introduction of systems that do look and sound human.

That possibility is just around the corner. Large language models like ChatGPT are already being used to power humanoid robots, such as the Ameca robots being developed by Engineered Arts in the U.K. *The Economist's* technology podcast, Babbage, recently conducted an interview with a ChatGPT-driven Ameca. The robot's responses, while occasionally a bit choppy, were uncanny.

Can companies be trusted to do the right thing?

The tendency to view machines as people and become attached to them, combined with machines being developed with humanlike features, points to real risks of psychological entanglement with technology.

The outlandish-sounding prospects of falling in love with robots, feeling a deep kinship with them or being politically manipulated by them are quickly materializing. I believe these trends highlight the need for strong guardrails to make sure that the technologies don't become politically and psychologically disastrous.

Unfortunately, technology companies cannot always be trusted to put up such guardrails. Many of them are still guided by Mark Zuckerberg's famous motto of moving fast and breaking things – a directive to release half-baked products and worry about the implications later. In the past decade, technology companies from Snapchat to Facebook have put profits over the mental health of their users or the integrity of democracies around the world.

When Kevin Roose checked with Microsoft about Sydney's meltdown, the company told him that he simply used the bot for too long and that the technology went haywire because it was designed for shorter interactions.

Similarly, the CEO of OpenAI, the company that developed ChatGPT, in a moment of breathtaking honesty, warned that 'it's a mistake to be relying on [it] for anything important right now … we have a lot of work to do on robustness and truthfulness.'

So how does it make sense to release a technology with ChatGPT's level of appeal – it's the fastest-growing consumer app ever made – when it is unreliable, and when it has no capacity to distinguish fact from fiction?

Large language models may prove useful as aids for writing and coding. They will probably revolutionize internet search. And, one day, responsibly combined with robotics, they may even have certain psychological benefits.

But they are also a potentially predatory technology that can easily take advantage of the human propensity to project personhood onto objects – a tendency amplified when those objects effectively mimic human traits.

17 March 2023

Is generative AI bad for the environment? A computer scientist explains the carbon footprint of ChatGPT and its cousins

An article from The Conversation.

By Kate Saenko, Associate Professor of Computer Science, Boston University

Generative AI is the hot new technology behind chatbots and image generators. But how hot is it making the planet?

As an AI researcher, I often worry about the energy costs of building artificial intelligence models. The more powerful the AI, the more energy it takes. What does the emergence of increasingly more powerful generative AI models mean for society's future carbon footprint?

'Generative' refers to the ability of an AI algorithm to produce complex data. The alternative is 'discriminative' AI, which chooses between a fixed number of options and produces just a single number. An example of a discriminative output is choosing whether to approve a loan application.

Generative AI can create much more complex outputs, such as a sentence, a paragraph, an image or even a short video. It has long been used in applications like smart speakers to generate audio responses, or in autocomplete to suggest a search query. However, it only recently gained the ability to generate humanlike language and realistic photos.

Using more power than ever

The exact energy cost of a single AI model is difficult to estimate, and includes the energy used to manufacture the computing equipment, create the model and use the model in production. In 2019, researchers found that creating a generative AI model called BERT with 110 million parameters consumed the energy of a round-trip transcontinental flight for one person. The number of parameters refers to the size of the model, with larger models generally being more skilled. Researchers estimated that creating the much larger GPT-3, which has 175 billion parameters, consumed 1,287 megawatt hours of electricity and generated 552 tons of carbon dioxide equivalent, the equivalent of 123 gasoline-powered passenger vehicles driven for one year. And that's just for getting the model ready to launch, before any consumers start using it.

Size is not the only predictor of carbon emissions. The open-access BLOOM model, developed by the BigScience project in France, is similar in size to GPT-3 but has a much lower carbon footprint, consuming 433 MWh of electricity in generating 30 tons of CO2eq. A study by Google found that for the same size, using a more efficient model architecture and processor and a greener data center can reduce the carbon footprint by 100 to 1,000 times.

Larger models do use more energy during their deployment. There is limited data on the carbon footprint of a single generative AI query, but some industry figures estimate it to be four to five times higher than that of a search engine query. As chatbots and image generators become more popular, and as Google and Microsoft incorporate AI language models into their search engines, the number of queries they receive each day could grow exponentially.

AI bots for search

A few years ago, not many people outside of research labs were using models like BERT or GPT. That changed on Nov. 30, 2022, when OpenAI released ChatGPT. According to the latest available data, ChatGPT had over 1.5 billion visits in March 2023. Microsoft incorporated ChatGPT into its search engine, Bing, and made it available to everyone on May 4, 2023. If chatbots become as popular as search engines, the energy costs of deploying the AIs could really add up. But AI assistants have many more uses than just search, such as writing documents, solving math problems and creating marketing campaigns.

Another problem is that AI models need to be continually updated. For example, ChatGPT was only trained on data from up to 2021, so it does not know about anything that happened since then. The carbon footprint of creating ChatGPT isn't public information, but it is likely much higher than that of GPT-3. If it had to be recreated on a regular basis to update its knowledge, the energy costs would grow even larger.

One upside is that asking a chatbot can be a more direct way to get information than using a search engine. Instead of getting a page full of links, you get a direct answer as you would from a human, assuming issues of accuracy are mitigated. Getting to the information quicker could potentially offset the increased energy use compared to a search engine.

Ways forward

The future is hard to predict, but large generative AI models are here to stay, and people will probably increasingly turn to them for information. For example, if a student needs help solving a math problem now, they ask a tutor or a friend, or consult a textbook. In the future, they will probably ask a chatbot. The same goes for other expert knowledge such as legal advice or medical expertise.

While a single large AI model is not going to ruin the environment, if a thousand companies develop slightly different AI bots for different purposes, each used by millions of customers, the energy use could become an issue. More research is needed to make generative AI more efficient. The good news is that AI can run on renewable energy. By bringing the computation to where green energy is more abundant, or scheduling computation for times of day when renewable energy is more available, emissions can be reduced by a factor of 30 to 40, compared to using a grid dominated by fossil fuels.

Finally, societal pressure may be helpful to encourage companies and research labs to publish the carbon footprints of their AI models, as some already do. In the future, perhaps consumers could even use this information to choose a 'greener' chatbot.

23 May 2023

Research

Try to find out some more information on the carbon footprint of an AI model. Which company has the lowest footprint? Do the companies offset their carbon emissions?

THE CONVERSATION

Turns out there's another problem with AI – its environmental toll

AI uses huge amounts of electricity and water to work, and the problem is only going to get worse – what can be done?

By Chris Stokel-Walker

Technology never exists in a vacuum, and the rise of cryptocurrency in the last two or three years shows that. While plenty of people were making extraordinary amounts of money from investing in bitcoin and its competitors, there was consternation about the impact those get-rich-quick speculators had on the environment.

Mining cryptocurrency was environmentally taxing. The core principle behind it was that you had to expend effort to get rich. To mint a bitcoin or another cryptocurrency, you had to first 'mine' it. Your computer would be tasked with completing complicated equations that, if successfully done, could create a new entry on to the blockchain.

People began working on an industrial scale, snapping up the high-powered computer chips, called GPUs (graphics processing units), that could mine for crypto faster than your off-the-shelf computer components at such pace that Goldman Sachs estimated 169 industries were affected by the 2022 chip shortage. And those computer chips required more electricity to power them; bitcoin mining alone uses more electricity than Norway and Ukraine combined.

The environmental cost of the crypto craze is still being tallied – including by the *Guardian* this April.

The AI environmental footprint

A booming part of tech – which uses the exact same GPUs as intensely, if not more so, than crypto mining – has got away with comparatively little scrutiny of its environmental impact. We are, of course, talking about the AI revolution.

Generative AI tools are powered by GPUs, which are complex computer chips able to handle the billions of calculations a second required to power the likes of ChatGPT and Google Bard. (Google uses its own similar technology, called tensor processing units, or TPUs.)

There should be more conversation about the environmental impact of AI, says Sasha Luccioni, a researcher in ethical and sustainable AI at Hugging Face, which has become the de facto conscience of the AI industry. (Meta recently released its Llama 2 open-source large language model through Hugging Face.)

'Fundamentally speaking, if you do want to save the planet with AI, you have to consider also the environmental footprint [of AI first],' she says. 'It doesn't make sense to burn a forest and then use AI to track deforestation.'

Counting the carbon cost

Luccioni is one of a number of researchers trying – with difficulty – to quantify AI's environmental impact. It's difficult for a number of reasons, among them that the companies behind the most popular tools, as well as the companies selling the chips that power them, aren't very willing to share details of how much energy their systems use.

There's also an intangibility to AI that stymies proper accounting of its environmental footprint. 'I think AI is not part of these pledges or initiatives, because people think it's not material, somehow,' she says. 'You can think of a computer or something that has a physical form, but AI is so ephemeral. Even for companies trying to make efforts, I don't typically see AI on the radar.'

That ephemerality also exists for end users. We know that we're causing harm to the planet when we turn on our cars because we can see or smell the fumes coming out of the exhaust after we turn the key. With AI, you can't see the cloud-based servers being queried, or the chips rifling through their memory to complete the processing tasks asked of it. For many, the huge volumes of water coursing through pipes inside data centres, deployed to keep the computers powering the AI tools cool, are invisible.

You just type in your query, wait a few seconds, then get a response. Where's the harm in that?

Putting numbers to the problem

Let's start with the water use. Training GPT-3 used 3.5 million litres of water through datacentre usage, according to one academic study, and that's provided it used more efficient US datacentres. If it was trained on Microsoft's datacentres in Asia, the water usage balloons to closer to 5m litres.

Prior to the integration of GPT-4 into ChatGPT, researchers estimated that the generative AI chatbot would use up 500ml of water – a standard-sized water bottle – every 20 questions and corresponding answers. And ChatGPT was only likely to get thirstier with the release of GPT-4, the researchers forecast.

Estimating energy use, and the resulting carbon footprint, is trickier. One third-party analysis by researchers estimated that training of GPT-3, a predecessor of ChatGPT, consumed 1,287 MWh, and led to emissions of more than 550 tonnes of carbon dioxide equivalent, similar to flying between New York and San Francisco on a return journey 550 times.

Reporting suggests GPT-4 is trained on around 570 times more parameters than GPT-3. That doesn't mean it uses 570 times more energy, of course – things get more efficient

Key Facts

- Training GPT-3 used by 3.5 million litres of water through datacentre usage.

- Researchers estimate that the generative AI chatbot, ChatGPT would use up 500ml of water – a standard-sized water bottle – every 20 questions and corresponding answers.

- Researchers estimated that training of GPT-3, a predecessor of ChatGPT, consumed 1,287 MWh, and led to emissions of more than 550 tonnes of carbon dioxide equivalent, similar to flying between New York and San Francisco on a return journey 550 times

– but it does suggest that things are getting more energy intensive, not less.

For better or for worse

Tech boffins are trying to find ways to maintain AI's intelligence without the huge energy use. But it's difficult. One recent study, published earlier this month, suggests that many of the workarounds already tabled end up trading off performance for environmental good.

It leaves the AI sector in an unenviable position. Users are already antsy about what they see as a worsening performance of generative AI tools like ChatGPT (whether that's just down to their perception or based in reality isn't yet certain).

Sacrificing performance to reduce ecological impact seems unlikely. But we need to rethink AI's use – and fast. Technology analysts Gartner believe that by 2025, unless a radical rethink takes place in how we develop AI systems to better account for their environmental impact, the energy consumption of AI tools will be greater than that of the entire human workforce. By 2030, machine learning training and data storage could account for 3.5% of all global electricity consumption. Pre-AI revolution, datacentres used up 1% of all the world's electricity demand in any given year.

So what should we do? Treating AI more like cryptocurrency – with an increased awareness of its harmful environmental impacts, alongside awe at its seemingly magical powers of deduction – would be a start.

1 August 2023

Britons think artificial intelligence will cost jobs… but not their own

Six in ten expect more jobs will be lost to robotics and AI than will be created.

By Matthew Smith

BT will reportedly be shedding up to 55,000 jobs by the end of the decade with up to a fifth replaced by technologies including artificial intelligence.

The results of a new YouGov poll show that almost two thirds of Britons (64%) believe 'more jobs will be lost to automation by robotics/AI than will be created', with a mere 7% expecting they will create more opportunities than they close down. One in eight (12%) expect numbers will remain about the same, while 17% are unsure.

Among workers themselves, 62% expect more jobs to be lost than gained. Yet when they are asked whether jobs like their own will primarily be done by humans or by robots or AI 30 years from now, the majority (59%) still see a human future. Only a quarter (25%) expect their line of work to become dominated by machines.

Likewise, few workers say they are worried about the impact that robotics or artificial intelligence will have on their current job (14%) or their future career (22%).

Younger workers – who have much more career still to get through – are unsurprisingly more likely to be worried than those who are on the cusp of aging out of working life. Even so, 18-24 year old workers are still more likely to be unworried (48%) than worried (36%), including only 7% who are 'very worried'.

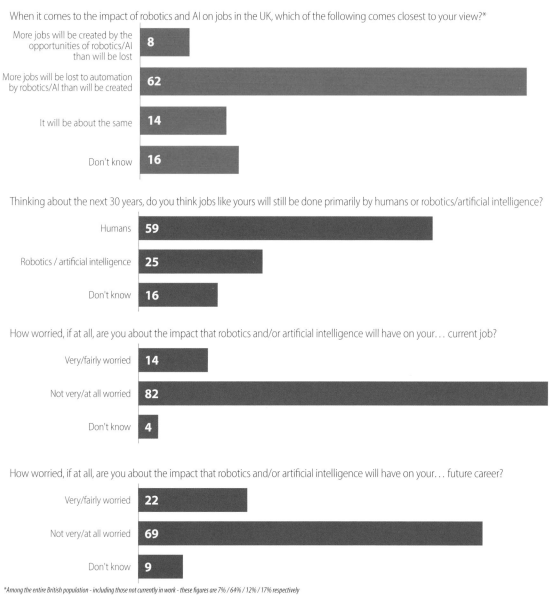

Britons think AI will cost jobs... but not their own
% of 1,169 working Britons

When it comes to the impact of robotics and AI on jobs in the UK, which of the following comes closest to your view?*

More jobs will be created by the opportunities of robotics/AI than will be lost	8
More jobs will be lost to automation by robotics/AI than will be created	62
It will be about the same	14
Don't know	16

Thinking about the next 30 years, do you think jobs like yours will still be done primarily by humans or robotics/artificial intelligence?

Humans	59
Robotics / artificial intelligence	25
Don't know	16

How worried, if at all, are you about the impact that robotics and/or artificial intelligence will have on your… current job?

Very/fairly worried	14
Not very/at all worried	82
Don't know	4

How worried, if at all, are you about the impact that robotics and/or artificial intelligence will have on your… future career?

Very/fairly worried	22
Not very/at all worried	69
Don't know	9

*Among the entire British population - including those not currently in work - these figures are 7% / 64% / 12% / 17% respectively
Source: YouGov

How many Britons think they understand about artificial intelligence?

% of each group who say they have a "great deal" or "fair amount" of understanding about each topic

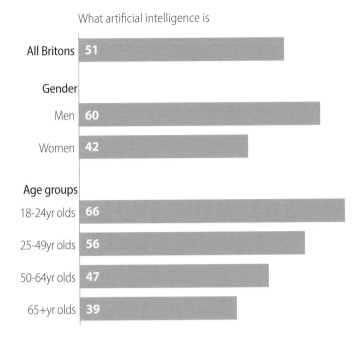

What artificial intelligence is

All Britons	51
Gender	
Men	60
Women	42
Age groups	
18-24yr olds	66
25-49yr olds	56
50-64yr olds	47
65+yr olds	39

The issues surrounding artificial intelligence

All Britons	41
Men	52
Women	33
18-24yr olds	55
25-49yr olds	47
50-64yr olds	38
65+yr olds	32

Source: YouGov

Among those workers who specifically think that more jobs will be lost to automation than will be created, 65% say they aren't worried about the impact on their own career.

How much do Britons think they understand about AI?

Half of Britons say they either know a great deal (7%) or a fair amount (44%) about what artificial intelligence is, while 41% say they don't know very much and 5% say they know nothing at all.

Slightly fewer people (41%) claim to know much about 'the issues surrounding artificial intelligence'.

As is generally the case in opinion polls, men are more likely than women to claim greater knowledge, and in this case younger people are more likely to feel they have a grip on the topic than their elders.

When it comes to the overall impact that artificial intelligence will have, Britons are more likely to be pessimistic (35%) than optimistic (19%), with a further 34% saying they are neither optimistic or pessimistic.

The more familiar a person feels they are with the issues surrounding AI, the more likely they are to have either optimistic or pessimistic expectations for the technology – although the optimistic side grows far more rapidly with awareness. While only 5% of those who confess no understanding of the issues around AI say they are optimistic about the tech, fully 28% have a negative view. Among those who consider themselves to have a great deal of understanding, there is a 35pt increase to 40% with an optimistic view, compared to a 15pt increase to 43% holding a negative view.

How does perceived familiarity with the issues surrounding AI affect optimism about the technology?

Would you say you are optimistic or pessimistic about the impact that artificial intelligence will have overall? %

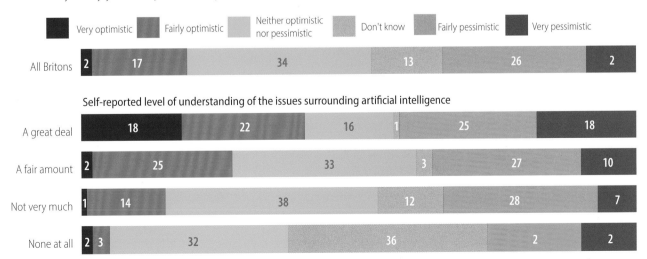

Legend: Very optimistic | Fairly optimistic | Neither optimistic nor pessimistic | Don't know | Fairly pessimistic | Very pessimistic

	Very optimistic	Fairly optimistic	Neither optimistic nor pessimistic	Don't know	Fairly pessimistic	Very pessimistic
All Britons	2	17	34	13	26	2

Self-reported level of understanding of the issues surrounding artificial intelligence

	Very optimistic	Fairly optimistic	Neither optimistic nor pessimistic	Don't know	Fairly pessimistic	Very pessimistic
A great deal	18	22	16	1	25	18
A fair amount	2	25	33	3	27	10
Not very much	1	14	38	12	28	7
None at all	2	3	32	36	2	2

Source: YouGov

Most Britons have little to no confidence that tech companies will develop AI responsibly, or that government will effectively regulate it

How much confidence do you have, if at all, in the following: %

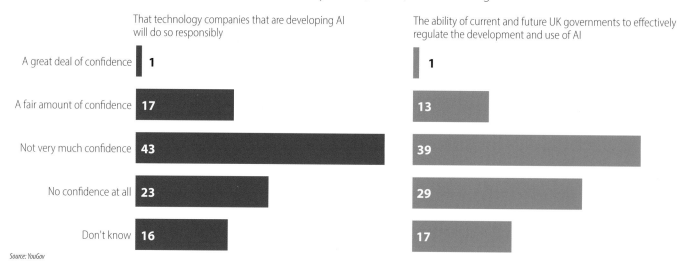

That technology companies that are developing AI will do so responsibly

A great deal of confidence	1
A fair amount of confidence	17
Not very much confidence	43
No confidence at all	23
Don't know	16

The ability of current and future UK governments to effectively regulate the development and use of AI

A great deal of confidence	1
A fair amount of confidence	13
Not very much confidence	39
No confidence at all	29
Don't know	17

Source: YouGov

Britons who are optimistic about the impact of AI tend to think that technology companies are going to develop it responsibly

How much confidence do you have, if at all, in the following: %

That technology companies that are developing AI will do so responsibly

	Optimistic about AI	Neither optimistic nor pessimistic about AI	Pessimistic about AI
A great deal of confidence	6	0	0
A fair amount of confidence	46	18	5
Not very much confidence	37	50	47
No confidence at all	6	15	44
Don't know	4	18	4

The ability of current and future UK governments to effectively regulate the development and use of AI

	Optimistic	Neither	Pessimistic
A great deal of confidence	5	1	1
A fair amount of confidence	31	15	4
Not very much confidence	43	45	40
No confidence at all	14	21	50
Don't know	7	18	5

Source: YouGov

Most Britons lack confidence in tech companies and government on AI

Earlier this week, the CEO of OpenAI (the firm which owns ChatGPT) told a US Congress hearing that AI could 'cause significant harm to the world', expressing willingness to work with lawmakers regarding his own company's technology, and made a number of proposals for how to regulate artificial intelligence.

The British public is, however, sceptical that those able to influence the future of AI can be trusted to do so.

Two thirds of Britons say they have little to no confidence that technology companies that are developing AI will do so responsibly (66%), or in the ability of current and future UK governments to effectively regulate the development and use of AI (68%).

Expectations are dramatically different depending on how optimistic or pessimistic people are about AI. Fully 52% of those who are optimistic about the impact artificial intelligence will have say they have a great deal or fair amount of confidence in tech companies to develop the technology responsibly – although far fewer have the same level of faith in the government's ability to regulate it effectively (36%).

Among those with pessimistic expectations for AI, just 5% have much confidence in either group.

19 May 2023

Chatbots sometimes make things up. Not everyone thinks AI's hallucination problem is fixable

Spend enough time with ChatGPT and other artificial intelligence chatbots and it doesn't take long for them to spout falsehoods.

By Matt O'Brien

Spend enough time with ChatGPT and other artificial intelligence chatbots and it doesn't take long for them to spout falsehoods.

Described as hallucination, confabulation or just plain making things up, it's now a problem for every business, organization and high school student trying to get a generative AI system to compose documents and get work done. Some are using it on tasks with the potential for high-stakes consequences, from psychotherapy to researching and writing legal briefs.

'I don't think that there's any model today that that doesn't suffer from some hallucination,' said Daniela Amodei, co-founder and president of Anthropic, maker of the chatbot Claude 2.

'They're really just sort of designed to predict the next word,' Amodei said. 'And so there will be some rate at which the model does that inaccurately.'

Anthropic, ChatGPT-maker OpenAI and other major developers of AI systems known as large language models say they're working to make them more truthful.

How long that will take – and whether they will ever be good enough to, say, safely dole out medical advice – remains to be seen.

'This isn't fixable,' said Emily Bender, a linguistics professor and director of the University of Washington's Computational Linguistics Laboratory. 'It's inherent in the mismatch between the technology and the proposed use cases.'

A lot is riding on the reliability of generative AI technology. The McKinsey Global Institute projects it will add the equivalent of $2.6 trillion to $4.4 trillion to the global economy. Chatbots are only one part of that frenzy, which also includes technology that can generate new images, video, music and computer code. Nearly all of the tools include some language component.

Google is already pitching a news-writing AI product to news organizations, for which accuracy is paramount. The Associated Press is also exploring use of the technology as part of a partnership with OpenAI, which is paying to use part of AP's text archive to improve its AI systems.

In partnership with India's hotel management institutes, computer scientist Ganesh Bagler has been working for years to get AI systems, including a ChatGPT precursor, to invent recipes for South Asian cuisines, such as novel versions of rice-based biryani. A single 'hallucinated' ingredient could be the difference between a tasty and inedible meal.

When Sam Altman, the CEO of OpenAI, visited India in June, the professor at the Indraprastha Institute of Information Technology Delhi had some pointed questions.

'I guess hallucinations in ChatGPT are still acceptable, but when a recipe comes out hallucinating, it becomes a serious problem,' Bagler said, standing up in a crowded campus auditorium to address Altman on the New Delhi stop of the U.S. tech executive's world tour.

'What's your take on it?' Bagler eventually asked.

Altman expressed optimism, if not an outright commitment.

'I think we will get the hallucination problem to a much, much better place,' Altman said. 'I think it will take us a year and a half, two years. Something like that. But at that point we won't still talk about these. There's a balance between creativity and perfect accuracy, and the model will need to learn when you want one or the other.'

But for some experts who have studied the technology, such as University of Washington linguist Bender, those improvements won't be enough.

Bender describes a language model as a system for 'modelling the likelihood of different strings of word forms,' given some written data it's been trained upon.

It's how spell checkers are able to detect when you've typed the wrong word. It also helps power automatic translation and transcription services, 'smoothing the output to look more like typical text in the target language,' Bender said. Many people rely on a version of this technology whenever they use the 'autocomplete' feature when composing text messages or emails.

The latest crop of chatbots such as ChatGPT, Claude 2 or Google's Bard try to take that to the next level, by generating entire new passages of text, but Bender said they're still just repeatedly selecting the most plausible next word in a string.

When used to generate text, language models 'are designed to make things up. That's all they do,' Bender said. They are good at mimicking forms of writing, such as legal contracts, television scripts or sonnets.

'But since they only ever make things up, when the text they have extruded happens to be interpretable as something we deem correct, that is by chance,' Bender said. 'Even if they can be tuned to be right more of the time, they will still have failure modes – and likely the failures will be in the cases where it's harder for a person reading the text to notice, because they are more obscure.'

Those errors are not a huge problem for the marketing firms that have been turning to Jasper AI for help writing pitches, said the company's president, Shane Orlick.

'Hallucinations are actually an added bonus,' Orlick said. 'We have customers all the time that tell us how it came up with ideas – how Jasper created takes on stories or angles that they would have never thought of themselves.'

The Texas-based startup works with partners like OpenAI, Anthropic, Google or Facebook parent Meta to offer its customers a smorgasbord of AI language models tailored to their needs. For someone concerned about accuracy, it might offer up Anthropic's model, while someone concerned with the security of their proprietary source data might get a different model, Orlick said.

Orlick said he knows hallucinations won't be easily fixed. He's counting on companies like Google, which he says must have a 'really high standard of factual content' for its search engine, to put a lot of energy and resources into solutions.

'I think they have to fix this problem,' Orlick said. 'They've got to address this. So I don't know if it's ever going to be perfect, but it'll probably just continue to get better and better over time.'

Techno-optimists, including Microsoft co-founder Bill Gates, have been forecasting a rosy outlook.

'I'm optimistic that, over time, AI models can be taught to distinguish fact from fiction,' Gates said in a July blog post detailing his thoughts on AI's societal risks.

He cited a 2022 paper from OpenAI as an example of 'promising work on this front'.

But even Altman, at least for now, doesn't count on the models to be truthful.

'I probably trust the answers that come out of ChatGPT the least of anybody on Earth,' Altman told the crowd at Bagler's university, to laughter.

1 August 2023

Research

Do some research on the inaccuracy of AI. Can you find any examples where AI has provided incorrect information? What issues can arise from this?

Research

As a class, come up with a prompt for a chatbot to find out some information. Then, each of you should ask a variety of chatbots and see if you all get similar answers.

Have any of you been given incorrect answers?

ChatGPT isn't the death of homework – just an opportunity for schools to do things differently

An article from The Conversation.

By Andy Phippen, Professor of IT Ethics and Digital Rights, Bournemouth University

ChatGPT, the artificial intelligence (AI) platform launched by research company Open AI, can write an essay in response to a short prompt. It can perform mathematical equations – and show its working.

ChatGPT is a generative AI system: an algorithm that can generate new content from existing bodies of documents, images or audio when prompted with a description or question. It's unsurprising concerns have emerged that young people are using ChatGPT and similar technology as a shortcut when doing their homework.

But banning students from using ChatGPT, or expecting teachers to scour homework for its use, would be shortsighted. Education has adapted to – and embraced – online technology for decades. The approach to generative AI should be no different.

The UK government has launched a consultation on the use of generative AI in education, following the publication of initial guidance on how schools might make best use of this technology.

In general, the advice is progressive and acknowledged the potential benefits of using these tools. It suggests that AI tools may have value in reducing teacher workload when producing teaching resources, marking, and in administrative tasks. But the guidance also states:

Schools and colleges may wish to review homework policies, to consider the approach to homework and other forms of unsupervised study as necessary to account for the availability of generative AI.

While little practical advice is offered on how to do this, the suggestion is that schools and colleges should consider the potential for cheating when students are using these tools.

Nothing new

Past research on student cheating suggested that students' techniques were sophisticated and that they felt remorseful only if caught. They cheated because it was easy, especially with new online technologies.

But this research wasn't investigating students' use of Chat GPT or any kind of generative AI. It was conducted over 20 years ago, part of a body of literature that emerged at the turn of the century around the potential harm newly emerging internet search engines could do to student writing, homework and assessment.

We can look at past research to track the entry of new technologies into the classroom – and to infer the varying concerns about their use. In the 1990s, research explored the impact word processors might have on child literacy. It found that students writing on computers were more collaborative and focused on the task. In the 1970s, there were questions on the effect electronic calculators might have on children's maths abilities.

In 2023, it would seem ludicrous to state that a child could not use a calculator, word processor or search engine in a homework task or piece of coursework. But the suspicion of new technology remains. It clouds the reality that emerging digital tools can be effective in supporting learning and developing crucial critical thinking and life skills.

Get on board

Punitive approaches and threats of detection make the use of such tools covert. A far more progressive position would be for teachers to embrace these technologies, learn how they work, and make this part of teaching on digital literacy, misinformation and critical thinking. This, in my experience, is what young people want from education on digital technology.

Children should learn the difference between acknowledging the use of these tools and claiming the work as their own. They should also learn whether – or not – to trust the information provided to them on the internet.

The educational charity SWGfL, of which I am a trustee, has recently launched an AI hub which provides further guidance on how to use these new tools in school settings. The charity also runs Project Evolve, a toolkit containing a large number of teaching resources around managing online information, which will help in these classroom discussions.

I expect to see generative AI tools being merged, eventually, into mainstream learning. Saying 'do not use search engines' for an assignment is now ridiculous. The same might be said in the future about prohibitions on using generative AI.

Perhaps the homework that teachers set will be different. But as with search engines, word processors and calculators, schools are not going to be able to ignore their rapid advance. It is far better to embrace and adapt to change, rather than resisting (and failing to stop) it.

27 July 2023

Consider...

Do you think that teachers should stop or change homework to avoid the use of AI-generated answers in assignments? Write a persuasive letter or email to your teacher to give your views on AI-generated essays.

THE CONVERSATION

How the education sector can safeguard students using generative AI

The emergence of free-to-access generative AI programmes has the potential to transform the classroom in ways we're only beginning to understand; how can we protect students from the dangers of AI.

By Mel Parker

The internet, perhaps the last major technological innovation that's comparable to AI, changed the way that students communicate, work, and access information. But tools like ChatGPT appear to be doing much more than that. They're changing the way that students live, work, and interact with technology every day.

Just as with the internet, it's difficult to predict now what this technology may lead to in a decade's time. What is clear is students have been very quick to utilise these tools to support their school work. Recently commissioned research from RM Technology reveals that over two thirds of school students are using the technology frequently, with some using it to solve mathematics problems, write English essays, translate texts in foreign languages, and more.

Despite the huge opportunities AI tools offer, many students are also struggling with its rapid introduction into their education. This is an aspect of the technology that too few are talking about. And we'll need to do exactly that if we're to guide children towards using the technology efficiently and safely, and protect them from the dangers which come from others misusing AI.

Ultimately, anyone with a stable internet connection and a computer is now able to produce harmful multimedia content, which could have a severe impact on a student's wellbeing.

Providing 'safety rails' is key

While the majority of school children are using AI to help with their school work, its popularisation has also increased their anxiety about education. Seventy per cent are now worrying that they'll struggle in exams without the support of AI tools. Although 68% of students say they are now achieving better grades since they started using it, at least half of those children are experiencing guilt about doing so.

It's clear that there's a significant lack of guidelines for the use of AI in schools. Definite 'safety rails' that describe effective practice could help to dispel anxiety and reduce guilt. They would ensure that, in whatever way students use AI, it will not hamper their potential in exams.

Introducing effective safety rails, though, is going to present a considerable challenge to the sector. The technology is advancing at a pace that regulators struggle to match. The Online Safety Bill now passing through the House of Lords may soon be outdated.

Legislation and guidance take time and experience to draft. With AI, we have little of either.

The public and private sectors need to come together to ensure that schoolchildren in the UK are protected from the potential harms of this new technology.

Parents and teachers must be equipped so they can work together to safeguard students.

Government regulation will, without a doubt, prove crucial in guiding students' use of AI in schools. Nevertheless, other stakeholders will need to play their part, since effective regulation remains a faint reality at some point in the distant future.

That's where teachers come in. Yet, it can be difficult without any guidance or training for teachers to effectively guide students' use of AI. Indeed, just over a third of teachers believe that students have a better understanding of AI than they do.

Schools must work with their technology partners to implement comprehensive training plans that cover every aspect of the use of AI, both in the classroom and outside it. Teachers must be equipped with the knowledge to decide how and where it is appropriate and beneficial, and how to spot when a student is misusing AI.

Safeguarding outside the classroom

Outside the classroom, it becomes difficult for teachers to guide and protect their students while they use AI. But parents will play an important role, since teachers can only do so much once a child has left the classroom.

Communication between parents and teachers is therefore vital if students are to be safeguarded outside the classroom. Informative material around the schools policy on the use of AI and how parents can safeguard their children is a great example of how schools can work with parents to ensure students continue to have a safe experience with AI outside the classroom.

It is clear that the education sector needs to introduce guidelines for students' use of AI – and quickly. What we must also recognise is that government regulation may take a long time before it can offer that guidance and protection. Teachers, parents, and the private sector all have an important role to play. And through effective training and collaboration, they can ensure children can continue to learn in a safe and productive environment.

This piece was written and provided by Mel Parker, Educational Technologist at RM Technology.

6 September 2023

What if Artificial Intelligence saves the planet?

People instinctively assume that AI will lead to catastrophe. But what will the world look like if we get it right?

By Henry Shevlin

We live in an era of intellectual pessimism. While the left laments collapsing biodiversity, climate catastrophe, soaring inequality, and the exploitative logic of capitalism, the right has become increasingly fixated on rapid cultural and demographic change, immigration and the decline of traditional family structures. In the face of such gloom, the optimist risks seeming naive, or even callously indifferent towards the many problems we face. And where once technology was seen as a potential solution to social and economic challenges, now it is more likely to be seen as their source, in the form of divisive social media, distracting smartphones, or the harm done by extreme or misleading online content.

While governments are torn between these competing negative narratives – particularly when it comes to Artificial Intelligence – there is a curious dearth of voices offering a more positive vision. Tech companies may well be tripping over themselves with breathless press releases, and social media has quickly filled up with hype-merchants promoting AI as if it were a get-rich-quick scheme. But serious, constructive and positive commentary is thin on the ground.

Even so, there are some academics, intellectuals and policymakers who are engaged with the question of what a positive outcome may look like in our development and use of AI; Max Tegmark's Future of Life Institute, for example, ran a competition earlier this year dedicated to mapping out positive futures. However, in the current intellectual climate, optimism is being drowned out by shrill negative judgements. While there is a place for voices of caution – Jonathan Freedland certainly provided that in a recent Guardian piece headlined 'The future of AI is chilling' – it would be a mistake to regard AI solely as a source of harm and threat, a new techno-pathogen against which we must inoculate ourselves.

With an eye on history, we should bear in mind that techno-panics can exact a heavy toll, albeit one that manifests more as missed opportunities than as concrete catastrophes. Perhaps the most famous such missed opportunity in the west is that of nuclear power. Haunted by the ghosts of Chernobyl, Three Mile Island and Fukushima, we have largely given up on construction of new nuclear plants, in many cases opting for the less visible but more pernicious negative effects of fossil fuels. More consequential for many in the developing world, perhaps, is the case of Golden Rice, a genetically modified crop capable of saving millions of Vitamin A-deficient children from starvation. Its widespread adoption remains stalled, all progress now entangled in a byzantine web of compliance and uncertainty.

The lesson is not that we should adopt a laissez-faire attitude to new technologies; in the case of both nuclear power and genetically modified crops there are serious risks to be managed. But caution is not cost-free. Even as we congratulate ourselves for harms averted, we should not ignore the invisible graveyards filled with those who could have benefited from opportunities we declined to take.

The new wave of powerful AI systems represents the first truly radical advance of the 21st century. The first glimpse many people had of this new technological wave came in the form of OpenAI's ChatGPT, released in November last year. This is one of a new flurry of systems known as Large Language Models, which are statistical distillations of billions of pages of text and conversation.

Capable of answering scientific or technical questions, summarising articles, providing advice, writing code, or even sustaining a casual conversation, they have caught the popular imagination. According to one industry survey, almost a third of workers in white-collar professions are already making use of ChatGPT at work. The majority have not yet informed their employers.

But this new wave of AI systems did not spring from a technological vacuum, even though it may have appeared that way. Instead it is the first widely available commercial manifestation of developments that have been under way for over a decade. As a research field, AI has had 'summers' and 'winters': periods of intense hype and investment, followed by disillusionment and a drop in funding. The current AI summer dates back around 15 years and has come about through increasingly affordable processor power. There have been signs that a technological revolution was coming, such as the advances in image categorisation and machine translation, but until quite recently, even a reasonably well-informed observer of technological trends could have missed their significance.

Though ChatGPT may have been more evolutionary than revolutionary, its astonishing popularity – reaching a million users in less than a month – marked a shift in public awareness of what companies like OpenAI and DeepMind were actually doing. While businesses, policymakers, and the public were struggling to come to terms with what these tools could achieve, the response from much of the scientific and academic commentariat was much more negative.

One trend in the negative reaction to AI is represented by AI sceptics such as Timnit Gebru and Margaret Mitchell, computer scientists and AI ethicists who were formerly employed by Google. Both have been trenchant critics of Large Language Models, which they say merely feed back to us a probabilistic cocktail of our own utterances, famously describing them as 'Stochastic Parrots'. Far from being revolutionary, these systems, they suggest, have a conservative or even reactionary character, embodying and concentrating existing systematic patterns of discrimination, exploitation and oppression.

In contrast to these more socio-political concerns, a quite different and far more existential source of unease about AI came to greater public attention in March this year when more than a thousand prominent figures, including Elon Musk, and scientists such as Yoshua Bengio, Max Tegmark and Gary Marcus, signed a petition demanding a slow-down of AI research. Their worry was motivated less by the political and societal implications of large models, but by disquiet about our ability to control AI and ensure it correctly aligns with our goals. Indeed, some people in AI see ChatGPT's capabilities as a nascent threat to human existence itself.

Gebru, Mitchell and others say the signatories of this petition are succumbing to 'fearmongering and AI hype,' and accuse them of steering public debate towards imaginary risks and ignoring 'ongoing harms'. These harms include worries about surveillance, the erosion of privacy, the use and abuse of large datasets, and injustices such as algorithmic bias.

These ideas are a long way from the thought that we might one day lose control of AI or that it might threaten our very existence. Terrifying though that idea may sound, it also seems rather fanciful – as if a word processor could suddenly decide to topple the British government. After all, aren't AIs just another kind of device? Couldn't we simply program them to be benign? And if all else fails, couldn't we just pull the plug? Matters, alas, are not quite so simple.

To start with, modern AI systems are not hand-coded in the manner of computer programs of old. Instead, the role of human programmers is largely to develop learning algorithms and curate training data, with AI systems being an emergent product, akin to a microbiologist seeding petri dishes with bacterial cultures. While these systems are subsequently fine-tuned, they nonetheless often display unexpected or undesired attributes, as demonstrated earlier this year when Microsoft's GPT-powered Bing search engine advised a journalist to divorce his wife, while expressing a desire to steal nuclear codes and create a deadly virus.

Looking back over the last 60 years, we see that AI has gone from a novelty to an arguably transformative technology, with one human ability after another being equalled or surpassed by artificial intelligence, whether in games like chess and 'Go' or in verbal benchmarks such as the Graduate Record Examination used in the US and Canada – GPT-4 ranks in the top 10% of human test-takers.

And while AI systems were once highly specialised, the current wave of large AI models are far more general in their capabilities, equally capable of conversing, coding and creating images. More worryingly still, they also display a growing talent for manipulation and persuasion; Open AI's GPT-4, for example, can pass a wide range of social cognition tasks involving inferring subtle motives in agents, while Facebook's CICERO system is able to play the famously Machiavellian game of 'Diplomacy' at a competitive human level.

Given that these systems' cognitive abilities are continuing to improve rapidly, we cannot dismiss the idea that, one day, we will be outfoxed by a 'misaligned' AI system. To borrow an analogy from the AI safety literature, consider the case of a child who has inherited a large fortune and who must appoint a trustworthy executor to manage their affairs. Without appropriate legal safeguards, they are vulnerable to manipulation and deception from unscrupulous aides. Many young medieval monarchs fell foul of their advisers precisely because of such an asymmetry in knowledge, guile, and ruthlessness.

As soon as we build AIs that are more cunning and sophisticated than ourselves, we may find ourselves in a similar position. The simple answer as to why we cannot turn off a malevolent AI system, then, is that any superintelligence worthy of the name will not tip its hand so easily; by the time we realise anything is amiss, humanity's fate will have been sealed, whether subtly through subversion of our political systems, or more bluntly through traditional science-fiction scenarios like seizing control of our nuclear weapons or unleashing lethal genetic diseases under the guise of a cure for cancer.

Regardless of whether we fear AI for its near-term political and social harm or for the longer-term risks of existential catastrophe, we have important reasons to take the risks of AI seriously. Consequently, it would be reasonable to expect the emergence of concrete goals, for example greater regulation, transparency and accountability for big tech. But the different tribes who are alarmed by the rise of AI – from Silicon Valley libertarians on the one hand, to progressives and campaigners for social justice on the other – make such temporary truces challenging.

It is increasingly clear to me as a researcher in this field that the current wave of generative AI is not just the latest iteration in a cycle of hype and disappointment, but a significant moment in the history of our species. In the last five years alone, generative AI has moved faster than even the most fevered optimists expected, as can readily be seen in the improvements in image models.

It is not hard to see areas in which AI can have a positive impact even in the near-term. Healthcare, for example, is an industry that faces constantly increasing costs and staff shortages, and as many developed economies experience a rapidly growing elderly population and a shrinking workforce, this will only increase. In principle, by using biometric data gathered from smartphones and wearable devices, AI could provide profound benefits, allowing early detection of cancers, the better monitoring of infectious disease and drug discoveries. Researchers in the US have recently used AI to identify a new antibiotic named abaucin, which could prove effective at killing bacteria immune to most current drugs. AI-powered language models could make basic diagnostic medicine more accessible, especially in parts of the world where healthcare is currently unaffordable for many people.

Careful society-wide discussions will need to be had about how to protect core values, such as patient confidentiality, and how to mitigate the risk of AIs making diagnostic errors. But we must remember that excessive caution or slowness to adopt these new tools carries a cost.

AI also has potentially huge applications in education, not just in making it more affordable but also more accessible and more equitable. In 1984, the educational psychologist Benjamin Bloom found that one-to-one tuition vastly improved students' performance as compared to those taught in large group settings. This widely replicated finding has become known as Bloom's 2-Sigma Problem, and is problematic only insofar as it was assumed that one-to-one tuition could not be provided to every student (merely those whose parents could afford it). This constraint may cease to apply in future. Enterprising teachers are already finding ways to use generative AI in the classroom, with

services such as Math-GPT beginning to approximate the kind of guidance provided by a personal tutor.

There are again costs and harms to be carefully weighed in deploying AI in education, from short-term worries about the misuse of ChatGPT for cheating in coursework, to potential longer-term risks, such as the decline of writing skills. But a careful reckoning of these costs and benefits requires us to balance culturally dominant techno pessimism with open-minded imagination about the better future that might be open to us.

These are just two examples of fields where even the near-term benefits of artificial intelligence are clear. There are doubtless many further applications, as yet undreamed of. Tools like ChatGPT may seem like relatively mundane clerical assistants for now, but generative AI goes beyond gimmicks: it is a general-purpose technology, and in the coming decade we will find any number of powerful ways to apply it. Whether the technological revolution going on around us will rival the agricultural and industrial revolutions of previous eras in scale and impact remains to be seen – but there can be little doubt that the world a decade from now will be a very different place.

But that is no reason to sink into fatalistic pessimism or techno determinism – the idea that technology will inevitably assert its influence over human affairs. Collectively, we have the power to decide how we choose to deploy AI and what safeguards to put in place.

Here, it seems to me, there is a vital role to be played by our artists, writers, and academics in energising the collective public imagination not merely with visions of dystopias to be avoided, but with positive visions towards which we can strive. There is no shortage of narratives informing us of the potential dire consequences of AI, from the dystopian near-future visions of Black Mirror to the AI apocalypses of The Terminator and The Matrix. But what might the world look like if we get AI right? That, it seems to me, is the far more important question.

Dr Henry Shevlin is Associate Director of the Leverhulme Centre for the Future of Intelligence at the University of Cambridge, where he also serves as Director of the Kinds of Intelligence Programme.

31 May 2023

Who owns the song you wrote with AI? An expert explains

By Douglas Broom

- Artificial intelligence potentially empowers us all to become creators – but who owns the outcome?

- The World Economic Forum has warned that copyright laws need to change to keep up with the potential of AI.

- Here, a professor of technology discusses how AI is disrupting our ideas – and laws – around intellectual property.

Artificial intelligence offers even the least musical among us the chance to get in touch with our inner songwriter. But what happens if you create a hit? Who owns the copyright? And what about the artist whose style is being plundered to create the AI hit? It's a question troubling lawyers and digital media experts.

Which is why we sat down with New York University's Professor Arun Sundararajan at the World Economic Forum's 2023 Annual Meeting of the New Champions in Tianjin, China, to seek answers to a question that's far from theoretical.

'Generative AI systems don't just generate new content in abstraction, they can be tailored to generate content in the style of a specific person,' Professor Sundararajan explained. 'You can create new Beatles songs. You can write a poem like Maya Angelou.

Which is exactly what happened recently when a series of cover versions of popular songs started to appear on TikTok. It quickly emerged that the artists performing the tracks had never recorded them – they had been entirely produced using artificial intelligence (AI).

'What's the point of having intellectual property law if it can't protect the most important intellectual property … your creative process?' Sundararajan asked.

The Forum's 2023 Presidio Recommendations on Responsible Generative AI warned it was 'essential for policy-makers and legislators to re-examine and update copyright laws to enable appropriate attribution, and ethical and legal reuse of existing content.'

Here's a summary of our discussion with Arun Sundararajan, Harold Price Professor at New York University Stern School of Business.

Who owns what we create with AI?

Sundararajan: This is one of the central policy and consumer protection issues when thinking about the governance of generative AI. Generative AI systems don't just generate new content in abstraction, they can be tailored to generate content in the style of a specific person.

You can create new Beatles songs. You can write a poem like Maya Angelou. And, you know, at this point, the ownership of a person over their creative process, over their intelligence, starts to get challenged by technology.

One way to think about this is: 'What's the point of having intellectual property law if it can't protect the most important intellectual property – your individual intelligence, your creative process?'

In the past, we've never really asked this question because it was assumed by default that, of course, you own how you create things. And so at this point in time, we have to extend intellectual property law to protect not just individual creations, but an individual's process of creation itself.

Is it fair to say our identities are at stake?

Sundararajan: I think one's creative process is part of one's identity as a human, but it's also an important part of one's human capital. You know, you can spend decades becoming really good at doing things in a specific way, and you have an incentive to do that because you own it and because you can enjoy the spoils, the returns from all of that investment.

The trouble is that now a generative AI system can take hundreds of examples of what an individual has created and start to replicate their creative process, in some way stripping away their human identity or taking part of their human capital away from them.

And this is something that we are seeing a lot in the creative industries, in the art industry and the music industry. Cartoonists' styles are being replicated by art-generating AI. Musicians' styles are being replicated by music-generating AI.

Are there implications for business?

Sundararajan: It's not just an issue for creative artists. Someone could be really skilled at business development in a company. That talent, that know-how, those years of experience that are rendered into email exchanges with clients, a particular style of talking to a particular potential client, a particular sequence of messages, a particular sequence of phone calls that leads to you closing the deal...

And when this business development executive leaves the company, of course they leave their work product behind. But now this work product can be used to create a digital replica of them that they may not have intended to leave behind, but which they can't take with them the way we have always taken our human capital with us if we move jobs.

What is the difference between mimicry and replicating something human?

Sundararajan: Once something is successful, people start to mimic that style. Because if I'm a musician and I want to imitate someone's style, there will still be differences between my style and theirs that reflect my talent and my creative ability. With an AI twin, in some sense you are creating an exact replica rather than simply mimicking the style.

I think the second big difference is the scalability of this. Once this is encoded into an AI system, new creations can be generated at a breathtaking pace.

Who is most at risk from this?

Sundararajan: If you're an incredibly famous artist there's very little danger, because if someone generates a replica of your music, you can simply say 'It's not mine' and then it won't be as popular. You've got the brand that allows you to get the economic returns from your creations.

On the other hand, if you're an up-and-coming band that isn't very well-known and you start to do somewhat well on Spotify and someone encodes into AI your style of music and starts to generate, hundreds of different examples, then your ability to even build that brand can be curtailed before you had the chance to do it. And so you're unable to get the economic returns associated with either your talent or with your human capital.

So where does the law stand at the moment?

Sundararajan: Every AI system we use was created by training it on examples. A lot of discussion has focused on whether it's OK for AI companies to use other people's creations for this. The law is unclear at present.

Some people argue that this falls under the fair use doctrine, which exists in the US, the EU and China (although it may be called different things in different places) where if you're transforming what you're using into something sufficiently different in a way that won't affect the commercial value of what you're using, then it's OK, you're not infringing on copyright.

The law is also unclear today on the ownership associated with something generated by AI, a particular creation. So if an AI system writes a story or generates a piece of art or composes a song, in some jurisdictions, if it is completely AI-generated with no human participation at all, nobody owns it, it's in the public domain.

If there's enough human assistance, such as providing a storyline which the AI completes for you or you outline a song and the song is then AI-generated, then you can continue to own the copyright.

On the issue of who owns the creative process, there seems to be little or no law that is giving us a definitive answer on how we can reclaim ownership of our creative process.

What's likely to happen is the US, the EU or China – one of these three – is [going] to take a leadership role and define the first set of guidelines and laws around individuals' ownership of their creative processes and the use of data to train something like ChatGPT.

Why is the question of who owns your intelligence so important?

Sundararajan: The biggest difference between generative AI and the AI that preceded it is that generative AI can create entirely new content based on past patterns or examples. This means that it can create new content in the style of a particular person. So for the first time, the ownership of a particular individual's creative process or their intelligence is up for grabs.

And if we don't retain ownership of our intelligence and creative process, then individuals will have a much lower incentive to develop that intelligence, or that human capital, in the first place. And this is really bad for a capitalist society.

Ceding our creative process to an AI poses a challenge to creators – it takes away their identity, it gives them less incentive to develop the ability to practice their art. Some people might argue that's OK because, for the rest of the world, AI systems are going to generate a much greater variety of art, music and literature.

I think the jury is still out on which of these two camps is going to be right. But certainly being able to embed someone's creative process into a generative AI system takes away a bit of their humanity, a bit of their identity.

How do you see the future?

Sundararajan: Well, in the broadest sense, we need to retain ownership of our intelligence as humans. How do we update intellectual property law to protect not just individual creations, but to protect an individual's creative process?

There are some promising early steps towards drawing the line between AI creation and human creation. In many jurisdictions, something that is generated by an AI system has to be marked as such. And if you are interacting with an AI that appears like a human, you have to be told that you are interacting with an AI.

I think a logical next step would be to ask the question: 'If an AI is generating these artefacts, these objects, these pieces of art, who owns the creations of the AI?' And then the step after that is to decide who owns the creative process if that creative process mimics a particular individual.

18 August 2023

AI can replicate human creativity in two key ways – but falls apart when asked to produce something truly new

An article from The Conversation.

By Chloe Preece, Associate Professor in Marketing, ESCP Business School and Hafize Çelik, PhD candidate in management, University of Bath

I s computational creativity possible? The recent hype around generative artificial intelligence (AI) tools such as ChatGPT, Midjourney, Dall-E and many others, raises new questions about whether creativity is a uniquely human skill. Some recent and remarkable milestones of generative AI foster this question:

An AI artwork, The Portrait of Edmond de Belamy, sold for $432,500, nearly 45 times its high estimate, by the auction house Christie's in 2018. The artwork was created by a generative adversarial network that was fed a data set of 15,000 portraits covering six centuries.

Music producers such as Grammy-nominee Alex Da Kid, have collaborated with AI (in this case IBM's Watson) to churn out hits and inform their creative process.

In the cases above, a human is still at the helm, curating the AI's output according to their own vision and thereby retaining the authorship of the piece. Yet, AI image generator Dall-E, for example, can produce novel output on any theme you wish within seconds. Through diffusion, whereby huge datasets are scraped together to train the AI, generative AI tools can now transpose written phrases into novel pictures or improvise music in the style of any composer, devising new content that resembles the training data but isn't identical. Authorship in this case is perhaps more complex. Is it the algorithm? The thousands of artists whose work has been scraped to produce the image? The prompter who successfully describes the style, reference, subject matter, lighting, point of view and even emotion evoked? To answer these questions, we must return to an age-old question.

What is creativity?

According to Margaret Boden, there are three types of creativity: combinational, exploratory, and transformational creativity. Combinational creativity combines familiar ideas together. Exploratory creativity generates new ideas by exploring 'structured conceptual spaces,' that is, tweaking an accepted style of thinking by exploring its contents, boundaries and potential. Both of these types of creativity are not a million miles from generative AI's algorithmic production of art; creating novel works in the same style as millions of others in the training data, a 'synthetic creativity.' Transformational creativity, however, means generating ideas beyond existing structures and styles to create something entirely original; this is at the heart of current debates around AI in terms of fair use and copyright – very

much uncharted legal waters, so we will have to wait and see what the courts decide.

The key characteristic of AI's creative processes is that the current computational creativity is systematic, not impulsive, as its human counterpart can often be. It is programmed to process information in a certain way to achieve particular results predictably, albeit in often unexpected ways. In fact, this is perhaps the most significant difference between artists and AI: while artists are self- and product-driven, AI is very much consumer-centric and market-driven – we only get the art we ask for, which is not perhaps, what we need.

So far, generative AI seems to work best with human partners and, perhaps then, the synthetic creativity of the AI is a catalyst to push our human creativity, augmenting human creativity rather than producing it. As is often the case, the

hype around these tools as disruptive forces outstrips the reality. In fact, art history shows us that technology has rarely directly displaced humans from work they wanted to do. Think of the camera, for example, which was feared due to its power to put portrait painters out of business. What are the business implications for the use of synthetic creativity by AI, then?

Synthetic art for business

Synthetic creativity on demand, as currently generated by AI, is certainly a boon to business and marketing. Recent examples include:

- AI-enhanced advertising: Ogilvy Paris used Dall-E to create an AI version of Vermeer's The Milkmaid for Nestle yoghurts.

- AI-designed furniture: Kartell, Philippe Starck and Autodesk collaborated with AI to create the first chair designed using AI for sustainable manufacturing.

- AI-augmented fashion styling: Stitch Fix utilised AI to capture personalised visualisations of clothing based on requested customer preferences such as colour, fabric and style.

The potential use scenarios are endless and what they require is another form of creativity: curation. AI has been known to 'hallucinate' – an industry term for spewing nonsense –

and the decidedly human skill required is in sense-making, that is expressing concepts, ideas and truths, rather than just something that is pleasing to the senses. Curation is therefore needed to select and frame, or reframe, a unified and compelling vision.

31 May 2023

Consider...

The images featured on these pages were created by AI. Is there any way that you can tell that they are not actual artworks painted by a human? *(Hint... look at the hands)*.

Create

Using an AI generator, what artwork can you create? How detailed does your prompt need to be? Is the outcome as expected?

Will AI change culture forever? It already has

If AI can clone Kanye West and scoop a photography award, is it too late to be afraid? Baz Luhrmann, Rowan Williams and others weigh in.

By Tristram Fane Saunders

The June 1893 issue of *The Studio*, a London arts journal, was dedicated to an urgent debate: 'Is the Camera the Friend or Foe of Art?' If a machine can produce a beautiful, lifelike image at the touch of a button, does that devalue 'real' art? It's the same question artists are asking today, 130 years later, about artificial intelligence.

In the past few months alone, AI has altered every sphere of the arts in ways so swift and subtle, so varied and widespread, that keeping track of it all has been almost impossible. It is already clear that the impact of AI – with its rapidly developing ability to generate images, sounds and words – will be the defining change in this decade's cultural life. What is less certain is whether this change represents something to celebrate, or to fear.

Visual arts

Last month, the Creative category of the Sony World Photography Awards was won by a photo that wasn't a photo. As beautiful as it is eerie, *Pseudomnesia: The Electrician* is a sepia portrait of two women who never existed. Boris Eldagsen, a 53-year-old German photographer, made it using DALL-E 2, an AI tool launched a year ago that creates images in response to verbal prompts; a word paints a thousand pictures.

Eldagsen submitted his image to the contest 'as a cheeky monkey'. In doing so, he broke no rules, if only because nobody had thought to update them for the AI age. 'I expect photo competitions to do their homework,' he tells me. When he learnt he'd won, he wrote back to reject the prize, explaining how he'd made his 'photo'. Sony still listed him as the winner.

So he rented a tux and flew to London for the awards ceremony, where he hoped he'd have a chance to sound the alarm about AI art. (It is art, he says, but a new form – one that doesn't belong in photography contests.) When he realised he wouldn't be given an opportunity to deliver a speech, he climbed on stage and grabbed the microphone to explain why he was rejecting his prize. An awkward moment followed, then the show carried on as normal.

After a subsequent flurry of news coverage sparked a debate about AI art, the awards' organisers issued a statement referring to 'creators engaging with lens-based' work. It dodged the key point: Eldagsen had created his picture without using a lens at all.

Making AI images from text prompts – 'promptography', Eldagsen calls it – requires craft. It's like photography: sure, anyone could trigger the shutter on a camera at random and fluke a brilliant image. A true artist can do it again and again. Eldagsen's method involves 11 tiers of specific text prompts, covering every aspect of a composition. He teaches classes in it.

Creating a great picture still takes work, then, but making a mediocre one has never been easier. There's always been a gap between trained and amateur efforts, but with AI 'professionals are scared because that gap is becoming smaller', says Eldagsen. 'The crappy images of the past are going to disappear.'

For Kate Crawford, author of Atlas of AI, 'there's an issue that's more important in the background here'. When it comes to making AI images, 'the 'creativity', if you wish to call it that, is premised on an enormous capture of the commons'. In the 18th century, we lost our grazing land; in the 21st, we've lost our data.

AI image engines such as Midjourney and DALL-E 2 are trained by 'scraping' – an unpleasantly, but perhaps appropriately, violent word – millions of pictures and their captions from the internet. 'Every time you've uploaded a photograph of a friend on holiday, or engaged in any kind of online activity that is publicly visible, that is now harvested into these gigantic datasets,' Crawford tells me. 'People look at the output and say "Is this art?" rather than looking at the fundamental practice itself. Is this nonconsensual taking of everything the type of reality we want to promote?'

In January, cartoonist Sarah Andersen led a class-action lawsuit against (among others) the London-based company behind Stable Diffusion, an image creator that launched last August, which can be asked to make pictures in the style of any artist you like, including Andersen. Getty Images announced its own suit in February, arguing that Stable Diffusion had generated images that are near identical to its copyrighted photos, some even reproducing the watermarks that Getty adds to prevent plagiarism.

Stable Diffusion was trained on LAION-5B, a vast database of scraped images (the '5B' stands for five billion). Mass scraping without first seeking consent has, so far, been considered legal as 'fair use' in the US. Some people are unfazed by this. 'Personally, as an artist, I came to the conclusion that I can't protect the images I've created in my life,' says Eldagsen, breezily. But others feel queasy. Two artists, Mat Dryhurst and Holly Herndon, last year created a website called haveibeentrained.com, which you can use to check whether your pictures have been used in AI training sets. One woman discovered a photograph taken by her doctor for her private medical files had somehow been scraped by LAION-5B.

Since Stable Diffusion agreed to work with haveibeentrained. com to let people 'opt out', more than one billion pictures have been pulled from its training set. Progress, perhaps. But that 'grab everything first, then put some of it back if

you ask nicely' model isn't the only way of doing things. In March, Adobe released a rival AI image-maker, trained only on out-of-copyright pictures and those to which Adobe already owns the rights.

Yet, even if the training data are acquired legitimately, many artists concerned for their jobs would rather boycott AI. Much of the anxiety is focused on genres where an image's value as art is secondary to its use for something else: to sell a product, or share information, or brighten up a bit of text – areas where 'good enough' is more important than 'good'.

Illustrators are particularly uneasy. 'This is effectively the greatest art heist in history,' they warned this month in an open letter signed by hundreds. 'If you think this sounds alarmist, consider that AI-generated work has already been used for book covers and as editorial illustration, displacing illustrators from their livelihood… Generative AI art is vampirical, feasting on past generations of artwork even as it sucks the lifeblood from living artists.'

All the above complaints have been made about the relatively straightforward world of static pictures. But what about movies? Let's ignore screenwriters for the moment, since Hollywood usually does, and focus on the moving image itself, that flickering ghost on the silver screen. Can you trust what you see?

Cinema

Text-to-video AI had its Lumière moment last September, when Meta unveiled Make-A-Video, a digital tool that can generate short silent movies from a single sentence. Google announced a rival product, Imagen, a week later. Both are still only available to a handful of human testers, but, in March, open-access AI organisation Hugging Face made its own text-to-video tool, ModelScope, open to all. It soon went viral with a nightmarish film, prompted by the line 'Will Smith eating spaghetti', in which the actor appears to be gnawing fistfuls of pasta like a man possessed.

All good fun, but would a major director – such as Baz Luhrmann, whose recent *Elvis* biopic was nominated for eight Oscars – see any possibilities in AI? 'We're already using it,' he says. 'On *Elvis*, we used AI to blend Elvis's face with Austin [Butler, the lead actor]. I think it's something that, used in the right way, can be very helpful.'

Luhrmann believes AI assistants and avatars will one day be everywhere, and we'll stop cringing at them. It's just like mobile phones, he tells me over a martini after a recent event at the Design Museum in London. 'When they first came in, only plumbers and tradesmen used them. It was like, how gauche!' In time, AI 'will just feel human'.

But not too human. Speaking about AI on stage earlier that evening, Luhrmann had said: 'I'm not scared as a creative person… If I said [to an AI], "Write me a screenplay in the style of yours truly about King Lear", what it'll lack is a sense of humanity. It's the flaws, the imperfections that make us human.'

Yet there are flaws aplenty in *The Great Catspy*, an AI-generated film trailer that went viral 10 days later, the first I'd seen to combine video, music and speech. Its cast suffers an anomaly common among AI people: weird hands. (AI struggles to remember how many fingers the average human has – 12? 17?) But it is still impressive. The animation's jazz-age sheen, spoofing the glossy style of Luhrmann's own 2013 film of *The Great Gatsby*, is ghastly, yes, but it instantly makes Will Smith's pasta look prehistoric.

Hollywood has been using computer-generated imagery (CGI) for decades, but until recent leaps in AI, it has been a slow, expensive, labour-intensive business. When the British actor Henry Cavill had to reshoot scenes for 2017's *Justice League*, but couldn't shave the moustache he'd grown for another role, CGI was used to erase it at a reported cost of $25 million. With really advanced AI, that sort of tweak could be done on a shoestring in a single afternoon.

In January, a new sitcom, *Deep Fake Neighbour Wars*, proved we can now convincingly give any actor the face of, say, Idris Elba or Greta Thunberg on even a relatively measly British television budget. So, why not revive dead stars? 'I could be hit by a bus tomorrow and that's it – but performances can go on and on,' Tom Hanks said this month. Would audiences pay to watch a purely digital Hanks? 'There are some people that won't care, that won't make that delineation.'

Keanu Reeves, of course, has been fighting against *The Matrix* for decades. 'Early on, in the early 2000s or it might have been the 1990s, I had a performance [digitally] changed,' he told a reporter in February. 'They added a tear to my face, and I was just like, 'Huh?!' It was like, I don't even have to be here… When you give a performance in a film, you know you're going to be edited, but you're participating in that. If you go into deepfake land, it has none of your points of view.' Since then, Reeves requests a clause be added to his contracts stipulating that his performances won't be digitally manipulated.

Actors not protected by such a clause might be unsettled by the work of companies such as Flawless AI, whose AI-powered editing can effectively reduce them to ventriloquists' dummies. It'll prove handy for rereleasing films in international markets. Dubbed into another language? No problem, they can sync an actor's lips to new dialogue. Foul language? Nothing could be easier than replacing an F-word with 'fiddlesticks'. Need to swap a line criticising China with one praising the CCP? They could, in theory, do that, too. (Flawless AI declined *The Telegraph's* request for an interview.)

Visually manipulating an actor's mouth would be of little use if AI couldn't also adapt their voice. But AI can generate uncannily convincing human sounds. It speaks, sings, plays symphonies. You may have listened to AI music without even knowing it.

Music

'It's Christmastime, and you know what that means,' sang Frank Sinatra, woozily, in 2020. 'It's hot tub time!' Ol' Blue Eyes had been roused from his eternal slumber to sing a new ditty about festive bathing, via OpenAI's much-maligned Jukebox music generator.

Most of the samples from that now wobbly-looking experiment have since been wiped from SoundCloud by OpenAI, but bootlegs of the Sinatra AI survive as a reminder of just how much slicker soundalikes have become in a mere two years since.

This is not uncontroversial. The Human Artistry Campaign, launched in March by 40 major US music and entertainment groups, is set on protecting the 'likeness' of voices, among other rights. Tom Waits set a precedent in 1992 when he successfully sued Frito-Lay for more than $2 million, over a Doritos ad that mimicked his unique rust-and-bourbon timbre.

Yet protecting voices will be harder now that AI has made 'voice-cloning' effortless. Last week, Apple announced that, come the autumn, iPhones will come with voice-cloning capability. Just 15 minutes of vocal samples is all you'll need to make a person 'say' anything with text-to-speech. In January, Apple quietly started releasing audiobooks with AI narrators.

Cloned hip-hop is almost as easy as cloned conversation, as app designer and vlogger Roberto Nickson proved in March, turning his own voice into Kanye West's for a rap about anti-Semitism that went viral. In April, *Heart on My*

Sleeve, a credible AI banger sung 'by The Weeknd and Drake' was streamed nine million times before it was pulled for violating copyright.

In that same month, Jay-Z's agent fought to get an AI song copying his voice pulled from YouTube, while alt-pop star Grimes declared she'd happily split royalties with anyone who created a deepfake song with her voice; people have since taken her up on the challenge. (Grimes has long been ahead of the AI curve: in 2020 she created an AI lullaby for her and her ex-boyfriend Elon Musk's child.)

Musicians are divided. According to the French DJ David Guetta, 'the future of music is in AI', while for Black Eyed Peas' Will.i.am, it's 'a great co-pilot'. Yet to Nick Cave, it's 'a grotesque mockery' of humanity. 'Songs arrive out of suffering,' Cave declared in January. 'Data doesn't suffer.'

Nick Littlemore, of Australian pop band Pnau, sees AI's potential, but is more measured. AI won't replace humans, he tells me, 'but it could replace vocoders. I'm not a great singer, but [with AI] I could get any voice I want, mould it, and put it on a record.' He knows plenty about playing with stars' voices, too: Pnau's remix of an Elton John and Dua Lipa collaboration has had a billion streams. He also remixed Elvis's songs for Luhrmann's biopic. '[AI] would have been very useful for that,' he says. 'You certainly could clean his voice up now, even more so than three months ago.'

Thanks to AI, Littlemore now finds making videos a doddle. The faces of guest vocalists Bebe Rexha and Ozuna are the only 'real' sights amid a sea of AI animation in the band's latest music video, released last weekend. Another recent video 'was shot on an iPhone: no makeup, in two takes, at home… and it looks like a million bucks'.

But can AI make music that inspires an emotional response? Littlemore thinks about it. 'Probably. I mean, you hit the right chords in the right moments – it's all there in classical music. Take stuff like Bach…' There are, of course, conventions in Western classical music – common chord sequences, for instance – that can be described mathematically. Bach's The Well-Tempered Clavier, comprising pieces in all 24 minor and major keys, could serve as a blueprint for AI imitations.

In music, as in visual art, AI made it easier for amateurs to rustle up passable work quickly. This month, Google went public with MusicLM: write a sentence telling MusicLM what kind of track you'd like, and it'll compose it. Reviewers haven't been overly impressed; MusicLM is a late entry to an already flooded market.

One AI music service, Boomy, boasts its 'users have created 14,699,511 songs, around 14.04% of the world's recorded music'. Another, Aiva, has been formally recognised as a composer by the Society of Authors, Composers and Publishers of Music.

Again, like in the visual arts, the area of the music industry in which AI poses perhaps the biggest threat to human artists is one that's tucked away from the spotlight: background music. 'The majority of music we hear in our everyday life, even if we don't notice it, is background music,' says Tao Romera Martinez, COO of Japanese AI company Soundraw. 'Elevators, radio ads, TV ads, presentations, all those social media videos produced every day. It's a pretty huge market.'

If you want, say, three minutes of gothic synth and strings in E-minor, with an emotional climax at the 43-second mark, press a few buttons and Soundraw will generate it for you. In April, Universal Music Group sent an open letter to Apple Music and Spotify urging them not to let Universal's tracks be 'scraped' by AI. But Martinez is keen to distance Soundraw's methods from the mass-scrapers. 'We're training our AI model exclusively using music by our in-house music-producers,' he says.

It's quite possible you've heard Soundraw's creations. The company doesn't ask for attribution, so Martinez says he can never be '100 per cent certain' whether or not the background music on a video came from Soundraw. But they have 'lots' of subscribers 'from national TV channels – and not just in Japan. If they're paying money for the subscription, they probably are using it somewhere.'

All Soundraw's tracks are wordless instrumentals. When it comes to lyrics, AI is still pretty immature, says Pnau's Nick Littlemore. 'Now is not the time to be using it for something as deep and heavy as Nick Cave, but I think it can do Dr Seuss now. If we fast-forward 10 years and give AI a heroin habit, maybe it'll come out with William Blake?'

Literature

Tristram Fane Saunders, a cynical journalist, decided to write a short story using an AI text-generator. He started typing, and wasn't prepared for what happened next. As Tristram began to type, the words flowed effortlessly from his fingertips. He found himself lost in the story, and before he knew it, he had written several pages. But as he read over what he had written, he realised that something strange was happening. The story was taking on a life of its own, and Tristram felt as though he was no longer in control…

I wrote the first two sentences of that; the rest were coughed up in a few seconds by Sudowrite's free, no-frills AI fiction-writer. On screen, it is helpfully colour-coded: 'Any text Sudowrite writes is purple.' Ho ho. But its prose mostly isn't purple, in the sense of florid. It's fast, functional, pulpy. It reads like Dan Brown.

Sudowrite can't offer anything like the structural complexity of a novel. Or it couldn't until last week, when it launched Story Engine, a vastly more refined writing tool capable of mapping out chapter-by-chapter plot beats and character development. According to Sudowrite's founder James Yu, 'our awesome team worked with hundreds of novelists'. (Irate writers on Twitter were quick to brand the unnamed novelists 'scabs'.)

If Story Engine-assisted novels seem generic, will readers mind? Plenty of popular fiction is derivative, a new variation on a familiar story, and many of the novels in today's bestseller charts are there largely because of TikTok, where books are hashtagged to highlight commercially popular plot tropes (eg '#enemiestolovers').

If you haven't yet asked AI to write for you, your friends probably have. More than a billion people – one billion – have used AI text-generator ChatGPT since it was launched six months ago. Alternatives have subsequently appeared from Microsoft (Bing Chat) and Google (Bard), but ChatGPT remains the most popular. It was created with a Large Language Model (LLM) called GPT-3. An LLM's complexity

is measured in 'parameters'. GPT-3 has 0.175 trillion. GPT-4, launched in March, has 100 trillion. The human mind, for context, has been estimated at 100 trillion synapses. GPT-4 can 'read' and 'understand' a prompt-command of up to 20,000 words (or two-thirds the length of *Animal Farm*.)

As with Stable Diffusion and its Getty watermarks, ChatGPT has sometimes clung worryingly close to its training texts. In February, author Susie Alegre was alarmed to find ChatGPT had spat out whole paragraphs from her award-winning book *Freedom to Think*, uncredited.

AI has not yet written an acclaimed novel, but it has already generated millions of terrible stories. So many have been sent to Clarkesworld, America's leading sci-fi magazine, that in February it was forced to close submissions for the first time in its history.

As with art and music, it's creative writing put to commercial use, rather than literature for literature's sake, that's most vulnerable to the creep of AI. Don't fret about the novelist; worry about the guy who writes the words for the novel's back cover. This month, one book distributor revealed it was already using AI to help write the blurbs for its book jackets.

Screenwriters and script doctors are similarly at risk. When Hollywood's WGA union went on strike this month, among its demands was an agreement that 'AI can't write or rewrite literary material', that it 'can't be used as source material', and its writers' work 'can't be used to train AI'.

Then there's the literary form closest to any hack's heart: journalism. I don't mean broadcast news – although, incidentally, a Kuwaiti TV channel introduced a blonde AI newsreader last month. I mean the written word: essays and features, reportage and opinion, as elevated to an art by the likes of Hazlitt, Swift, Orwell and Amis. Will AI replace journalists? In some places, it already has.

Associated Press has been using AI to write stories (mainly business and sports) since 2014. Reach, owner of the *Daily Express* and *Daily Mirror*, began publishing AI-written articles in March, following BuzzFeed's lead. In April, a German magazine editor was sacked after running a fake AI-generated interview with former Formula One driver Michael Schumacher. This month, the *Irish Times* apologised for being tricked into running an opinion piece about fake tan that turned out to be written by ChatGPT.

Before he became a bestselling novelist, Neil Gaiman cut his teeth as a music journalist. He tells me he's now concerned about how AI has warped his old field. 'There's a whole new world waiting for us, of utterly convincing "facts" that are generated in things that are not actually sentences – they're just sentence-shaped.'

He gives a recent example. 'I'm a huge fan of Lou Reed. I picked up my phone, and there at the very top of the news app is 20 Best Songs of Lou Reed, Ranked. I thought, 'Oh great, that's something I'll spend five minutes on while making a cup of tea.' I read the description of the first song, and it wasn't right...' It seemed as if the 'journalist' had attempted to guess what the songs were about purely based on their titles. 'I realised this was an AI-generated article. The descriptions sound incredibly authoritative, unless you know anything at all about the songs.'

AI is changing the world too quickly for us to follow, says Gaiman. 'We're in a period that we haven't figured out yet, and by the time we do it'll be too late.'

Poetry is the oldest form of literature and, arguably, the most distilled expression of the human spirit. Don Paterson – one of the country's most respected poets – is running to be the next Oxford Professor of Poetry. Last month, in his election manifesto, he wrote that 'AI is the only serious technological challenge poetry has faced since Gutenberg.' So I asked him what he meant. Gutenberg's printing press, says Paterson, gave poetry 'a sense of its individual authorship that I don't think it suffered from until that point'. The individual suddenly mattered more than the communal, oral tradition.

If AI reduces our reverence for the author, shifting the focus from personality to technique, that might be a good thing, he says. It would undermine 'this sentimental delusion about inspiration as a completely inscrutable source. If you take a colder view, and look at the best effects in certain lines, some are amenable to analysis in ways that an AI could help you with.' The 'frisson' you get from a line of Shakespeare, he says, sometimes comes from the juxtaposition of words with close sounds but distant meanings (eg 'a little more than kin and less than kind'). With enough effort, you could code for that kind of wordplay.

If poets use AI 'like an enhanced personal assistant', is it really so different from a rhyming dictionary? 'I already have this stupid program I use that's able to cross-reference 40 thesauruses, because it's the kind of thing I have fun doing,' admits Paterson. 'But the interesting thing is: will the reader notice any difference if a poem has been composed with the help of AI or by traditional means?'

We may never have an AI Seamus Heaney, but less subtle writers are easier to imitate. 'If you're the sort of poet who writes the same poem again and again you may soon be literally predictable, especially if it's the kind of "ambient"poem that works through a series of images,' says Paterson. AI will soon 'be able to write a poem of the sort that would be eminently publishable – if that's any test of anything.'

When we realise how much popular, mediocre poetry could be convincingly imitated by machines, he says, there will be 'a crisis of the author. It will pose real questions about where the value lies; is it in authenticity? If it's not and it's in talent, what does it mean that certain talents appear to be easily imitable?' After all, creative talent is 'the last word in human agency. If that becomes imitable, it's a worry – it's like our souls have been stolen. And in a sense, they will have been.'

Rowan Williams – poet, philosopher, former archbishop of Canterbury – knows a thing or two about souls. He's been impressed by AI 'paintings and songs and poems, and I'm quite happy to say, yes, there is a beauty there, there is a shapeliness to that'. But, he tells me, they lack the depth of art. 'Sooner or later, someone's going to wake up to the fact there's no real resourcefulness there. You can produce a highly effective, highly efficient imitation of art, but then that's always been the case. It's important to remember that when AI does anything, anything at all, it's always imitating.' It just jams unlikely sources together, 'like a magazine competition that asks you to write a scene from Dostoevsky in the style of P. G. Wodehouse'. A mere mash-up is not creativity, he says.

Aidan Meller disagrees: 'I would argue that the combination of unlikely things is, effectively, what creativity is – it is creating something new.' Meller heads the team behind Ai-da, an AI android poet and painter who, in the past two years, has answered questions at the House of Lords, recited her own poetry to students at Oxford and exhibited her sculptures at the pyramids of Giza.

Ai-da takes in data through her camera-eyes and microphone-ears, processes them with her AI-brain, then uses an old-fashioned paintbrush and oils to fill a canvas with marks. Her paintings are derivative, insofar as they're based on everything she's heard, seen and thought. Are humans any different?

But perhaps trying to make AI resemble an individual human artist is missing the point. It's possible that great AI art won't sound like any of us, but like all of us. Boris Eldagsen thinks it already does: 'Because the training material was so widely collected from the internet, it's a mirror of the human condition – what Carl Jung would have called the collective unconscious.'

Last month, a giant screen in New York's Times Square was covered by an image that seemed a good symbol for human and machine working hand-in-hand: looping, handwritten ones and zeros, Cursive Binary, an artwork by artist and poet Sasha Stiles. She co-writes her poems with an AI she has fed with her own writing, and speaks of it almost like a friend. 'I personally have had many experiences reading computer-generated poetry where it's moved me deeply,' she tells me. 'I don't think it matters whether it's a human-written text or machine-written text. If it moves you, I think that's poetry.'

If experts can't consistently tell human and AI art apart – and they can't – does the AI art have just as much value? Does our response to it matter more than how it was made?

'It really matters that great poems get written – and it doesn't matter a damn who wrote them,' said Ezra Pound. Unless he didn't. The line is apocryphal, but it's a great line. Should it matter who – or what – wrote it?

If we really want to understand the AI future, says Williams, we should be studying art not by AI, but about AI. For him, *Klara and the Sun*, Kazuo Ishiguro's 2021 novel, told from the point of view of an 'artificial friend' android 'is such an interesting fantasy, because it tried to represent what the first-person stance of a highly developed artificial intelligence might be like. Do we treat it as if it had rights and claims? It's through that sort of imaginative projection that we begin to get a better grip on this than through looking at the kinds of tasks AI can fulfil.'

So, let's imagine an AI friend like Ishiguro's Klara. One without a human mind or soul, but with something sufficiently mind-like and soul-like to sing, and write, and paint in a way that stirs the soul in us. Would her existence devalue our own?

'That wouldn't to my mind undermine the idea of human distinctness or dignity,' says Williams. 'Because we would still be what we were. We'd still get wet when it rained, we'd still die, we'd still have sex, and enjoy meals, and do a variety of other things that machines don't seem to be interested in doing.'

27 May 2023

Useful Websites

Glossary

Algorithm

An algorithm in a set of instructions or rules that are used to solve a problem or perform a task.

Anthropomorphism

Anthropomorphism is the attribution of human traits, emotions, or intentions to non-human entities such as animals or inanimate objects.

Artificial intelligence (AI)

Artificial intelligence is the intelligence of machines or software, opposed to the intelligence of animals or humans.

Carbon footprint

A carbon footprint is a measure of the amount of carbon dioxide released into the atmosphere as a result of the activities or actions of a particular individual, organization, or community.

Carbon offsets

Carbon offsets are a reduction in greenhouse gas emissions made in order to compensate for greenhouse gas production somewhere else. Offsets can be purchased in order to comply with caps, such as the Kyoto Protocol. For example, rich industrialised countries may purchase carbon offsets from a developing country in order to satisfy environmental legislation.

Carbon tax

Carbon tax is a fee imposed on the production, distribution and burning of fossil fuels responsible for CO2 and other greenhouse gas emissions. It incentivises companies to cut their carbon emissions and invest in cleaner, greener options.

Chatbot

A chatbot is a computer program that simulates conversation with human users, often using generative AI to automate responses.

ChatGPT

ChatGPT is a text-based generative AI system, created by San Francisco-based firm OpenAI. Launched in November 2022, it has had over 100 million users in the six months following the launch.

Cognitive

A term referring to the mental processes of perception, memory, judgement and reasoning.

Cybercrime

Crime with some kind of 'computer' or 'cyber' aspect to it: using modern telecommunication networks such as the Internet (like chat rooms, e-mails and forums) and mobile phones (texting) to intentionally physically or mentally harm and cause distress. Computer viruses, cyberstalking, identity theft and phishing scams are some examples of cybercrime.

Data Protection Act 2018

The Data Protection Act controls how your personal information is used by organisations, businesses or the government. The Data Protection Act 2018 is the UK's implementation of the General Data Protection Regulation (GDPR).

Digital literacy

The ability to think critically about information consumed online, to be able to use the internet safely and with proficiency.

Digital native

A person who has grown up surrounded by digital technology, such as mobile phones, computers and the Internet (the current 12- to 18-year-old generation).

Internet

A worldwide system of interlinked computers, all communicating with each other via phone lines, satellite links, wireless networks and cable systems.

Internet of Things

This term refers to the network of objects that now connect via the Internet. For example, cars, watches, fridges, etc.

Legislation

A law or body of laws that aim to regulate behaviours or actions.

Online harms

Content or activity that can cause harm to internet users, particularly children and vulnerable people. These behaviours can harm people either emotionally or physically.

Virtual reality

Virtual reality is the creation of a virtual environment presented to our senses in such a way that we experience it as if we were really there. It uses computer modelling and simulation to create a three-dimensional visual environment.

Index